T0139353

Environmental Systems and Societies

FOR THE IB DIPLOMA

Jill Rutherford
Gillian Williams

OXFORD
UNIVERSITY PRESS

Contents

About the authors

Jill Rutherford has some 30 years of teaching, administrative and board experience within international schools. She was the founding director of the IB Diploma at Oakham School, England and has held senior examining positions at the IB. She is currently academic director of Ibicus International, which offers workshops to IB teachers around the world. She holds two degrees from the University of Oxford. Her passion lies in teaching and writing about the IB Environmental Systems and Societies course.

Gillian Williams graduated from Reading University and has taught Environmental Systems, Geography and TOK on the international circuit since 1993. In her international career Gillian has held various leadership positions including Deputy Head, Head of Year and Head of Department and has worked in Malaysia, Tanzania, Jordan and China. In 2011 she began advising on the IB Environmental Systems and Societies curriculum review. She is a workshop leader (online and face-to-face) and part of the IB Global Mentoring Team.

Practical work

An introduction to practical work in ESS

Well done for choosing to take the ESS (Environmental Systems and Societies) IB Diploma course. It is a great course which gets you thinking about and doing investigations into our environment.

The ESS guide says that you must carry out practical work.

Practical work is intended to:

- illustrate and reinforce concepts
- develop an appreciation of the hands-on nature of fieldwork
- develop an appreciation of the benefits and limitations of investigative methodologies.

You do practical work to gain the skills you will need for ESS.

Practical work should be totally integrated into the ESS course teaching.

Practical work in the ESS course is made up of two parts:

1. You must complete an investigation and submit it to the IB to gain an IB Diploma. It must be your own work and written up following the guidelines and criteria which are stated in the ESS guide. The Investigation – called the IA (Internal Assessment).

   This is an individual investigation which you choose.

   - It is marked by your teacher and moderated by the IB.

2. You must also complete a practical scheme of work which is part of your learning in the ESS course and should be integral to the theoretical teaching in your classes. The PSOW (**P**ractical **S**cheme **o**f **W**ork).

   In this part of the course you carry out investigations to help you understand the concepts involved. It is marked by your teacher(s) and may be required by the IB as well. Throughout the PSOW, you will encounter and practise important techniques which will be good preparation for the IA. It is marked by your teacher(s) but the investigations will not be submitted to the IB. A class copy of the PSOW form will be included with the moderation sample.

Check your understanding of the terms used in ESS practical work.
- IA
- PSOW
- Investigation
- Practical work
- Coursework
- Fieldwork
- Lab work

The PSOW:

- is the practical course that your teacher plans and your whole class carries out
- should take about 20 hours of the course, not including writing up time
- does not have a specific number of activities to be done
- reflects the breadth and depth of the syllabus so covers most of the topics
- contains some simple and short and some longer and more complex tasks
- is listed in the form ES&S/PSOW (which is the same for your whole class)
- is recorded by you in your PSOW file or lab book
- is not assessed by the IB but is assessed by your teacher(s).
- form is sent to IB for moderation purposes

Make sure you have and read these IB documents:
- IB ESS course guide.
- Poster: Are you completing your IB assignments honestly?
- IB animal experimentation guidelines.
These should be available from your teacher or Diploma Coordinator.
- Your school's IB academic honesty policy.

▲ **Figure 1.1** Doing practical work in the field – in this case, a pond

'In the field' is a term that scientists use to describe investigations and experiments done outside in an ecosystem. In your work it refers to any experiments you conduct in the laboratory, in the classroom or on a beach, on a mountainside or by a pond.

What you need to do

The PSOW could include a wide range of practical activities, such as:

- a short piece of lab work done in 30 minutes or less
- longer monitoring investigations over weeks or different seasons
- computer simulations
- making, developing or using models
- gathering information from questionnaires
- fieldwork
- data-analysing exercises
- researching and analysing databases for secondary data.

Practical work could be:

Methodologies

- Values and attitude surveys or questionnaires
- Interviews
- Issues-based inquiries to inform decision-making
- Observational fieldwork (natural experiments)
- Field manipulation experiments
- Ecosystem modelling (e.g. mesocosms or bottle experiments)
- Laboratory work
- Models of sustainability
- Use of systems diagrams or other valid holistic modelling approaches

- Elements of environmental impact assessments
- Secondary demographic, development and environmental data
- Collection of both qualitative and quantitative data

Analytical techniques

- Estimations of NPP/GPP or NSP/GSP
- Application of descriptive statistics (measures of spread and average)
- Application of inferential statistics (testing of null hypotheses)
- Other complex calculations
- Cartographic analysis
- Use of spreadsheets or databases
- Detailed calculations of footprints (e.g. ecological, carbon, water)

Rationale for practical work

Do you know these sayings?

'I hear and I forget. I see and I remember. I do and I understand.' – Confucius

不口不若口之，口之不若口之，口之不若知之，知之不若行之。学至于行之而止矣。Xunzi

Roughly translated as:

'Not hearing something is not as good as hearing it, hearing is not as good as seeing, seeing is not as good as knowing, knowing is not as good as acting; true learning continues until it is put into action.'

There is a general belief amongst educators that you, the student, should 'do' in order to understand and we suspect that you think so too. You probably remember the lessons where you were active in some way more than you remember others where you were only a receiver of information.

We need to think about the aims of the ESS course and the aims of the IB for you, the student, to understand why you need to do practical work in ESS.

The ESS course should give you a perspective on the complex relationships between humans and our environment. It wants you to explore these relationships using many tool kits and through different lenses – scientific, cultural, economic, ethical, political and social.

The course will be successful for you if it enables you to recognize and evaluate the impact that individuals and societies have on the world and its ecosystems. Once you can do that you are better equipped to make wise decisions about your actions in future. The course is aimed at getting you to think holistically about environmental issues and to solve problems both practically and theoretically using management solutions. There is no point in saying 'things are bad and getting worse' about the environment if we cannot then try to make them better. So this is an optimistic course because you can make things better.

The systems approach provides the core methodology of the ESS course. It is complemented by other influences, such as economic, historical, cultural, sociopolitical and scientific factors, to provide a holistic perspective on environmental issues. During the course, you will look at examples on a variety of scales, from local to global, and in an international context.

The aims of the ESS course are to enable you to:

1. acquire the knowledge and understanding of environmental systems at a variety of scales

2. apply the knowledge, methodologies and skills to analyse environmental systems and issues at a variety of scales

3. appreciate the dynamic interconnectedness between environmental systems and societies

4. value the combination of personal, local and global perspectives in making informed decisions and taking responsible actions on environmental issues

5. be critically aware that resources are finite, and that these could be inequitably distributed and exploited, and that management of these inequities is the key to sustainability

6. develop awareness of the diversity of environmental value systems

7. develop critical awareness that environmental problems are caused and solved by decisions made by individuals and societies that are based on different areas of knowledge

8. engage with the controversies that surround a variety of environmental issues

9. create innovative solutions to environmental issues by engaging actively in local and global contexts.

So what has this got to do with you carrying out practical work?

Aims 2 and 9 of ESS are particularly relevant to 'doing' rather than just hearing or seeing. Throughout the course, there are key skills that you need to develop in order to complete the IA task and do well in the final exams.

In the ESS guide, the practical skills are listed under the 'Applications and skills' headings of each topic and many of these are either practical skills themselves or can be understood by carrying out a practical activity. Here are a few examples:

Sub-topic 2.5

- Design and carry out ecological investigations.
- Construct simple identification keys for up to eight species.
- Evaluate sampling strategies.
- Evaluate methods to measure at least three abiotic factors in an ecosystem.
- Evaluate methods to investigate the change along an environmental gradient and the effect of a human impact in an ecosystem.
- Evaluate methods for estimating biomass at different trophic levels in an ecosystem.
- Evaluate methods for measuring or estimating populations of motile and non-motile organisms.
- Calculate and interpret data for species richness and diversity.
- Draw graphs to illustrate species diversity in a community over time, or between communities.

Sub-topic 4.2
- Evaluate the strategies that can be used to meet an increasing demand for freshwater.
- Discuss, with reference to a case study, how shared freshwater resources have given rise to international conflict.

Sub-topic 8.4
- Evaluate the application of carrying capacity to local and global human populations.
- Compare and contrast the differences in the ecological footprint of two countries.
- Evaluate how EVSs impact the ecological footprints of individuals or populations.

What the IB wants you to achieve

The IB assessment objectives tell you what will be assessed:

1. Demonstrate knowledge and understanding of relevant:
   - facts and concepts
   - methodologies and techniques
   - values and attitudes.

2. Apply this knowledge and understanding in the analysis of:
   - explanations, concepts and theories
   - data and models
   - case studies in unfamiliar contexts
   - arguments and value systems.

3. Evaluate, justify and synthesize, as appropriate:
   - explanations, theories and models
   - arguments and proposed solutions
   - methods of fieldwork and investigation
   - cultural viewpoints and value systems.

4. Engage with investigations of environmental and societal issues at the local and global level through:
   - evaluating the political, economic and social context of the issues
   - selecting and applying the appropriate research and practical skills necessary to carry out investigations
   - suggesting collaborative and innovative solutions that show awareness and respect for the cultural differences and value systems of others.

Format of the assesment

Assessment objective	Which component addresses this assessment objective?	How is the assessment objective addressed?
Objectives 1–3	Paper 1	Case study
Objectives 1–3	Paper 2	Section A: short answer questions Section B: two essays from a choice of four
Objectives 1–4	Internal assessment	Individual investigation assessed using markbands

The objectives will be tested in the examinations through the use of the command terms (provided in the appendix of the guide).

Assessment component	Weighting	Approximate weighting of objectives (%)		Duration (hours)
		1&2	3	
Paper 1 (case study)	25	50	50	1
Paper 2 (short answers and structured essays)	50	50	50	2
Internal assessment	25	Covers objectives 1, 2, 3 and 4		10

Being academically honest

You need to understand what the IB means by the terms:

- academic honesty
- malpractice
- plagiarism
- collusion.

Academic honesty in the International Baccalaureate (IB) is a principle informed by the attributes of the IB learner profile. In teaching, learning and assessment, academic honesty serves to promote personal integrity and engender respect for others and the integrity of their work. Upholding academic honesty also helps to ensure that all students have an equal opportunity to demonstrate the knowledge and skills they acquire during their studies.

Malpractice is behaviour that results in, or may result in, the candidate or any other candidate gaining an unfair advantage in one or more assessment components. Examples of malpractice:

- **Plagiarism** – the representation of the ideas or work of another person as your own.

- **Collusion** – supporting malpractice by another candidate, as in allowing your work to be copied or submitted for assessment by another candidate.

- **Duplication of work** – the presentation of the same work for different assessment components and/or IB diploma requirements.

Go through the ESS guide and make a separate list of all the points in the left-hand column under 'Knowledge and understanding' and 'Applications and skills' that you think you can learn by doing something active. The actions could be in the laboratory or fieldwork, doing an activity in a group or going on a visit somewhere.

- **Misconduct** during an examination, including the possession of unauthorized material.
- **Disclosing information** to another candidate, or receiving information from another candidate, about the content of an examination paper within 24 hours of sitting the examination.

All the coursework you submit to your teacher or the IB must be your own work and ideas. If you reference or mention the work or ideas of other people, you must cite these properly. If you do not, you may be guilty of plagiarism or collusion. These carry heavy punishments in the IB regulations. And beware, as the IB has no way of knowing whether you deliberately carried out academic misconduct or did it by mistake. Whatever your intent, the penalties are the same.

> Read these documents from the IB: *Academic honesty, The Diploma Programme: From principles into practice* and the relevant articles in *General regulations: Diploma Programme.*

All coursework – including work submitted for assessment – is to be authentic, based on the student's individual and original ideas with the ideas and work of others fully acknowledged. Assessment tasks that require teachers to provide guidance to students or that require students to work collaboratively must be completed in full compliance with the detailed guidelines provided by the IB for the relevant subjects.

You must make sure, and then sign a document to say you are sure that your IA work is your own and you have not plagiarized any of it or colluded with someone else in writing it. Once your work has been sent to the IB, you cannot then retract it; it is too late.

Your teacher must authenticate your work before it is sent to the IB. That means they must be sure it is your work and only your work. They may do this by looking at any or all of these:

- your initial proposal
- your first draft
- references cited
- writing style
- using a web-based plagiarism detection service, e.g. Turnitin.

A word on ethical practice in ESS

- No experiments involving other people will be undertaken without their written consent and their understanding of the nature of the experiment.
- No experiment will be undertaken that inflicts pain on, or causes distress to, humans or live animals, in line with the IB animal experimentation policy.
- No experiment or fieldwork will be undertaken that damages the environment.

2 Internal assessment (IA) – your investigation

The IA:

- is your own piece of work
- should take about 10 hours (not including writing up time)
- is assessed by the IB
- contributes up to 25% of your final mark.

So it is important to do as well as possible on this as it could make the difference of a couple of grades to you.

On pages 32 and 39 you can find two full IA samples.

Why do IA?

The reasons you do the internal assessment are to:

- demonstrate your skills and knowledge in a more relaxed setting than exams
- pursue a personal interest.

An additional bonus is that it constitutes 25% of the final ESS grade, so you have up to a quarter of your grade long before you sit your final exams.

The internal assessment task involves the completion of an individual investigation of an environmental systems and societies research question that you have designed yourself. The investigation should be submitted as a written report.

Note: Any investigation that is to be used for IA should be specifically designed to address the assessment criteria. You must therefore be provided with a copy of the assessment criteria.

If you undertake an extended essay in ESS, it should not be based on the same research question as the IA.

What does it involve?

For this element of your assessment you must complete an **individual** investigation that you have designed and implemented. You have to address the assessment criteria directly so make sure you have a copy of these before you start.

The time allocated for you to do your internal assessment is 10 hours.

Don't panic. That time includes time:

1. with your teacher to discuss the requirements of the internal assessment
2. discussing the IB animal experimentation policy – you are not allowed to design an investigation that will kill or distress animals
3. consulting with your teacher about your plan, your methods of data collection and anything else you are unclear on
4. doing the investigation
5. for consultation with your teacher about your progress
6. for feedback on one draft.

Things to remember

- The report must be between 1500 and 2250 words in length. Less than that and you have probably not dealt with the research question in enough depth; more than that and you will lose marks for the communication criterion and the examiner stops marking at 2250 words.

- Identify a particular aspect of ESS (something that interests you) that allows you to apply your findings to the wider environmental and societal context. Show the interconnection between the natural and human sciences – people and the environment.

- Pick an environmental **issue** so that you have the opportunity to discuss solutions either in the local or global context. This cannot be emphasized enough – marks hang on this one aspect.

- Develop sound methodologies that will give you enough data to analyse.

- ESS is an interdisciplinary subject that can take methodology and analytical techniques from both the natural and human sciences – be adventurous and check out the techniques given in this book and the ESS course companion or another ESS textbook.

- You can use qualitative and quantitative data, primary and/or secondary data, statistical analysis, mapping techniques etc. Make sure you develop a focused research question that comes from a broad area of environmental interest. If you do this you can apply your findings at the local or global level.

Warnings

- You cannot do your ESS internal assessment and your extended essay based on the same research question.
- The work must be your own work.
- It is not your teacher's job to tell you what to do.
- Teachers are required to authenticate your work – if they suspect any form of malpractice they will not do so. This will be done through
  - discussion with you
  - reference to your initial plans
  - checking a first draft
  - checking your style of writing
  - use of plagiarism detection tools.
- If the teacher misses any malpractice but the IB catches it, you may lose your diploma.

What makes a good IA?

The internal assessment is marked out of 30; you are assessed against six criteria. They are clearly set out in the ESS guide (2017). Make sure you have a copy of them. You cannot hope to achieve the top marks if you do not know what you are aiming for!

Criterion	Marks	Percentage
Identifying the context (CXT)	6	20
Planning (PI)	6	20
Results, analysis, conclusion (RAC)	6	20
Discussion and evaluation (DE)	6	20
Applications (A)	3	10
Communication (C)	3	10
TOTAL	30	100

▲ **Figure 2.1** IA criteria and marks

The IB uses markbands so your teachers use a best-fit approach to find the correct mark for your work, as not all indicators are present in all pieces of work.

In the criteria there are verbs called **command terms** that tell you what to do. You need to know what these mean.

Look up the definitions of the IA command terms (see p. 129).

Analyse	Identify
Construct	Interpret
Deduce	Justify
Describe	List
Design	Outline
Discuss	State
Evaluate	Suggest
Explain	

The criteria

Criterion	What you do	Notes	
CXT	State relevant research question (RQ).	Issue could be local or global.	✔
	Discuss relevant environmental issue in context of RQ.		✔
	Explain connection between the issue and the RQ.		✔
PI	Design repeatable method to investigate the RQ.	Make sure you collect enough relevant data (ask your teacher how much).	✔
	Justify your choice of sampling strategy.		✔
	Describe the risk assessment.		✔
	Describe the ethical considerations.		✔
RAC	Construct diagram, chart or graph of data.	Data can be quantitative (values) or qualitative (observations) or both. Display relevant patterns (if any).	✔
	Analyse data correctly.		✔
	Interpret trends, patterns or relationships in data.		✔
	Deduce a valid conclusion.		

DE	Evaluate the conclusion.	Evaluate in context of the environmental issue you talked about in CXT. Method modifications should cover one or more weaknesses that would have a large effect on the results.	✔
	Discuss strengths, weaknesses and limitations of your method.		✔
	Suggest modifications of your method.		✔
	Suggest further areas for research.		✔
A	Justify one potential application and/or solution to the issue discussed in CXT.	Base your application justification on what you found out in your research.	✔
	Evaluate relevant strengths, weaknesses and limitations of the solution.		✔
C	Structure and organize the report well.	You do not have many words for all this (1500–2250) so think and plan before you start writing.	✔
	Use consistent terminology.		✔
	Be concise.		✔
	Be logical and coherent.		✔
	Write a bibliography or cite sources if needed.		✔

▲ **Figure 2.2** Checklist for IA criteria

Identifying the context: worth 6 marks (20%)

This criterion looks at how well you establish and explore the environmental issue (local or global) and then use it to develop the research question. So how well does it all hang together?

- Make sure your research question is clear, significant and well focused.
- Show you understand the environmental issue and its relevance locally or globally.
- Explain how the research question is linked to the environmental issue you are investigating (background information to set the scene).

You will not get good marks if you:

- have a vague/broad research question like 'Is global warming fact or fiction?'
- just give brief notes about the environmental issue with nothing about its relevance
- forget to mention how the research question and environmental issue are linked.

Do not leave the moderator wondering what you are talking about.

Planning: worth 6 marks (20%)

In this criterion you will be assessed on the method of data collection and its relevance to your research question. It will look at whether or not you have assessed safety matters and the environmental and ethical considerations of the investigation.

- You must design a repeatable method that allows you to collect sufficient relevant data.
  - Repeatable means that you must explain it well enough for someone else to replicate. A cake recipe is no good if it cannot be followed well enough to produce the same cake.
  - Sufficient relevant data means enough data for you to analyse your research question. You do not want a load of data that is unrelated to your question.
- Make sure you justify the method, especially sampling strategies.
- Show you have assessed and understood the ethical and environmental considerations.

You will not get good marks if you:
- Design a method that
  - is so unclear it could not be repeated
  - generates insufficient data
  - generates plenty of data but does not address your research question.
- Forget to explain and justify your sampling strategy.
- Choose the wrong sampling strategy.
- Ignore risks, ethical and environmental factors.

Results, analysis and conclusion: worth 6 marks (20%)

This criterion is about how methodical you are when you collect, record, and process your data. It also looks at how you interpret the data and the conclusion you come to.

Remember:
- Data tables need
  - column and row headers
  - good titles explaining exactly what they are showing
  - units in the headers
  - consistent decimal places.
- Show how you did your calculations – one sample is good.
- Graphs need
  - labelled axes with units
  - good titles explaining exactly what they are showing
  - lines or curves of best fit, if appropriate
  - small data points – not massive blobs.

- Make sure you process your data and then present it in the most appropriate way to enhance analysis.
- If you have collected the data, present it – all of it, qualitative and quantitative.

- Describe and interpret all relevant patterns, trends or relationships in the data, including identifying and explaining anomalies (do not just ignore them!).
- Analyse the data and reach a valid conclusion.
- If you have a hypothesis, remember you have not proven or disproven it; you can only say whether you accept or reject it.

You will not get good marks if you:

- Do not process and present all your data or you have made some errors.
- Only describe the data you have presented and ignore patterns, trends and anomalies.
- Do not give a conclusion or you give a conclusion that is unsupported by the data you have collected.

Discussion and evaluation: worth 6 marks (20%)

This criterion looks at your discussion and conclusion and how well they are related to the environmental issue you proposed in the first place. It also looks at how you evaluate your investigation.

At this point you may fall foul of a poor starting point so think your investigation ALL the way through to the end before you start it. Remember:

- Your research question must be based in a broad environmental area of interest.
- This facilitates discussion of the broader implications of your findings (local or global).
- The broader discussion may not be directly linked to your findings – you will probably not have enough data for that but you can project and extrapolate. You can extend your thinking to link to the real world and discuss solutions linked to your environmental issue.

So, if you have not thought it through you will either lose marks or you will have to start again, neither of which are good!

- Evaluate your conclusion and make the links to the environmental issue VERY clear.
- Discuss fully the strengths, weaknesses and limitations of the method you used. DO NOT use things that you really should have thought of in the first place.
- Suggest how things could have been improved or even extended through further research.

You will not get good marks if you:

- Only describe your data or make no link to the environmental issue.
- Give a half-hearted evaluation of some of the strengths and/or weaknesses and/or limitations.
- List superficial modifications and/or extensions with no discussion.

Applications: worth 3 marks (10%)

Here you will be assessed on the identification and evaluation of the applicability of the outcomes of your investigation. How does what you have discovered apply to the environmental issue you first identified? So again, if you did not identify an issue you cannot score on this criterion.

- You must use your findings to propose and justify a possible application/solution to the environmental issue you identified in the investigation.
- Evaluate the strengths, weaknesses and limitations of your solution/application.

You will not get good marks if you:

- Simply state an application/solution.
- Do not link your findings to the environmental issue.
- Do not justify your application/solution.
- Do not state some strengths, weaknesses and limitations of the solution/application.

Communication: worth 3 marks (10%)

This is a holistic criterion (not tied in to any one section of your report). This focuses on presentation, how well your report is written – does it have a clear, coherent structure that is easy to follow and makes sense?

- Present a well-organized report with side headings (it can be very useful to follow the criteria headings).
- Make use of the appropriate terminology.
- Do not waffle – you can't really afford to waste words with this word limit so be concise.
- The report must be logical – hence the comment about sticking to the order of the criteria.
- You must be coherent – the person reading your report should be able to follow your line of thought easily. DO NOT JUMP ABOUT in your thinking.
- Give a clear bibliography that is complete – the system does not matter, just be consistent.

You will not get good marks if your report is:

- disorganized – jumps about between ideas and has no logical order
- lacking in the appropriate terminology
- disjointed and full of irrelevant information
- difficult to understand
- lacking a bibliography – this could cost you your diploma.

Sample IA investigations

ESS topics are interlinked so think about including more than one topic in your investigation. On the following pages, you will find some suggestions.

They are very specific and will only be applicable if you have the right circumstances in your local area. There are similar investigations which could be done and you can modify these suggestions to suit your area.

1. Environmental Impact Assessment (EIA)

Topics 1 and 2

Suggested research question:

What impacts will the new IB College building have on the local area (natural environment and human populations) and how can any negative impacts be mitigated?

If you are in an area where a new building is planned you have an ideal opportunity to conduct your own mini Environmental Impact Assessment (EIA). You may need to gain permission to conduct an initial survey – if it is within the school area that should not be a problem.

What is needed in an EIA?

There is no set way of conducting an EIA, but various countries have minimum expectations of what should be included. It is possible to break an assessment down into three main tasks:

- identifying impacts (scoping)
- predicting the scale of potential impacts
- limiting the effect of impacts to acceptable limits (mitigation).

Conducting your own EIA

In this case the study area is probably clearly defined, as an area will have been identified for the building. You will have to collect a wide range of data that will constitute a 'baseline study' before the project starts.

Baseline study (scoping)

1. You will need an accurate map of the designated building area. Google Maps or Google Earth can be useful here.
2. You will need to decide on a suitable sampling strategy – this will depend on whether the area is uniform or not:
   a. for a uniform area random sampling would be appropriate
   b. for an area where there is a possible environmental gradient present you will need to use transect lines.
3. Data you need to collect:
   a. climatic data – humidity, temperature, rainfall (using a rain gauge), wind speed, light intensity
   b. plant species abundance
      i. density
      ii. frequency
      iii. percentage coverage
      iv. species diversity
   c. animal species present
   d. amenity value of the area.

Techniques for all the above are described in Section 3 except for the last, amenity value.

Tip

- Make sure you have a focused research question and link it clearly to an appropriate environmental issue.
- Describe ALL your methods clearly and fully and justify your sampling techniques so that someone else could repeat it.
- Be creative in your data presentation – there are different data collection techniques that lend themselves creativity (but make sure they aid analysis).
- The EIA has a built-in solution step so make the most of it.

Amenity value

Amenity value means 'why do people like this place/thing and what does it add to their lives?' So, for instance, if you cut a tree down will that upset people because that tree adds to the overall feeling and look of the location? If you erect a building will it ruin the view?

This can be very hard to assess but a bipolar analysis could be used to get some indication of the value of the location.

Bipolar analysis of a picture

Use a picture of the area as it is now and ask people what they like about the area. You will have to adjust your questions to your location but here are some suggestions.

The area	+2	+1	−1	−2	The area
Is quiet and peaceful					Is very noisy and unpleasant
Has lots of natural vegetation that makes it nicer					Is unnatural and ugly
Has lots of birds and other wildlife					Has no wildlife
Is clean and free of litter					Is dirty and full of litter
Is useful: for playing sports, sitting and relaxing					Is not useful for anything
Looks and feels inviting and pleasant					Is not a nice place to be
Blends in with everything around it and adds to the overall quality of the whole area					Is different from everything around it

▲ **Figure 2.3** An example of a bipolar analysis

Conducting this simple bipolar analysis will enable you to give a numerical amenity value. You will have to be very careful if you conduct a survey like this because people may get worried about the new building project. It may be better for you to restrict your survey to the ESS class and friends.

This concludes your baseline study. You now know what you have before the new build starts.

Predicting the scale of potential impacts

This requires you to think about all the information you have collected and how the proposed building may impact each of them. Some additional research will be needed here.

There are two aspects of a new building project that will impact the area.

1. The building phase (excavation and construction)

    a. noise
    b. dust
    c. traffic congestion
    d. effect on utilities.

2. The final building.

    You have collected data on:

    a. climate
    b. plant species abundance
    c. animal species present
    d. amenity value of the area.

You job is to predict the impact of the building on these things.

Limiting the effect of impacts to acceptable limits (mitigation)

Now comes the problem-solving part – how can you reduce the impact of the building?

2. Investigate the changes in an area caused by human activity

Topic 2

Suggested research question:

What is the impact of footpath erosion on plant biodiversity?

▲ **Figure 2.4** Footpath erosion

> **Tip**
> - The environmental issue is obvious (footpath erosion) but don't forget to explain the local (or global) context.
> - Take time to describe your method so it is repeatable, justify your sampling strategies and discuss risks and ethical issues.
> - This is an easy one for solutions so do not forget to propose and discuss them fully.

This can be done in any area where people walk across natural vegetation – national parks, through woods, across fields, etc. It is also very common in schools with large green areas to cross. Think about what differences there may be in different areas – does it change as you go further away from a car park in a natural area? Is it worse in certain areas? Be original, think outside of the box.

1. This investigation needs transect lines as there is a definite change along a particular direction.

2. You will have to decide on your sampling interval and quadrat size etc.

3. You must do a minimum of three transects – more than three is better.

4. Keep moving away from the footpath (in both directions) until there is very little change in the data you are collecting.

5. Data you could collect (for methods see Section 3 and sub-topic 2.5 in the course companion):

   a. plant species abundance

      i. density

      ii. frequency

      iii. percentage coverage

      iv. species diversity (calculated using Simpson diversity index)

   b. infiltration rate of the soil

   c. amount of litter

   d. number of people that pass by.

These are just suggestions – you can probably think of others.

3. Investigate ecological footprints (secondary data followed by a questionnaire)

Topics 1 and 8

Suggested research question:

What is the relationship between ecological footprint and the level of development of a country and what are people's attitudes to possible solutions to high ecological footprints?

Part 1: Using secondary data look at the relationship between the ecological footprint and the level of development of a country. Level of development can be measured using:

- Human Development Index (HDI)
- Gross Domestic Product per capita (GDP US$)
- health care – life expectancy, infant mortality rate
- education – number of years in school, literacy rates
- or any other suitable measures.

Part 2: Once you have the data for the relationship, produce a questionnaire to assess:

- possible solutions
- people's attitudes to given solutions.

Part 1: Secondary data collection to establish the relationship between ecological footprint and level of development

There are a number of possible websites for secondary data – the following websites will provide a wide range of data.

- United Nations Development Programme (UNDP)
    - http://hdr.undp.org/en: home page with various useful tabs (data, countries, reports) and it is a great source for a wide range of data
    - http://hdr.undp.org/en/data: a list of all data available
- CIA world factbook
    - https://www.cia.gov/library/publications/the-world-factbook/ for a wide range of demographic and economic data – just select the country you are interested in.
- Gapminder
    - data tab for a range of data: http://www.gapminder.org/data/
    - Gapminder world tab for graphs: http://www.gapminder.org/world
- Nationmaster: http://www.nationmaster.com/index.php.

Websites specific to this investigation:

- Human Development Index (HDI)
    - The UNDP website is probably the source for most other websites for the HDI: https://data.undp.org/dataset/Human-Development-Index-HDI-value/8ruz-shxu – list of all countries and their HDIs
- Gross Domestic Product per capita (GDP $US)
    - UNDP: https://data.undp.org/dataset/GDP-per-capita-2005-PPP-/navj-mda7

Tip

This is a two-part investigation.

- Develop a sharply focused research question that is clearly linked to a specific environmental issue.
- Ecological footprints have both global and local significance but with a limited word count you may have to focus on one of them.
- Make sure you cover both parts equally as regarding planning.

Tip

- You should make sure that you have at least two, preferably three, sources of data for each of your variables. You can then calculate the mean for the various sources which will lessen the impact of rogue data points.
- You will have to decide on a sampling method to select a minimum of 30 countries for the study.
- Once you have collected all the data a scattergraph is probably the best method to show the relationship and a line of best fit will highlight the relationship.
- Statistical tests are also necessary to show the strength of the relationship.

- ■ World Bank: http://data.worldbank.org/indicator/NY.GDP.PCAP.CD
- ■ CIA world factbook: https://www.cia.gov/library/publications/the-world-factbook/rankorder/2004rank.html
- ■ Nationmaster
- ● Health care – life expectancy, infant mortality rate
  - ■ UNDP: https://data.undp.org/dataset/Table-7-Health/iv8b-7gbj
  - ■ Indexmundi: http://www.indexmundi.com/facts/topics/health
  - ■ CIA world factbook
- ● Education – number of years in school, literacy rates
  - ■ UNDP: https://data.undp.org/dataset/Table-8-Education/mvtz-nsye
  - ■ CIA world factbook
  - ■ Nationmaster
  - ■ Gapminder
- ● Ecological footprint
  - ■ Nationmaster
  - ■ http://en.wikipedia.org/wiki/List_of_countries_by_ecological_footprint, taken from Ecological Footprint Atlas 2010.

Part 2: Primary data collection about solutions/attitudes towards solutions

Generally speaking there is a relationship between ecological footprint and the level of development of a country, but it is not always that straightforward. Many of the MEDCs are beginning to recognize the problems of a high ecological footprint and they are introducing measures to reduce their footprint.

For this stage you can design a questionnaire that identifies some of the measures to reduce the ecological footprint of a country and conduct a questionnaire to establish which ones people are most likely to adopt. The ecological footprint is made up of a number of aspects and there are various ways to reduce the footprint of each aspect.

Aspect of footprint	Explanation	Measures to reduce it
Carbon emissions	Carbon emissions from burning fossil fuels and land required to absorb that	Buy food grown locally Use energy-efficient light bulbs Install solar panels Turn off lights Fly less Walk, cycle, use public transport
Crop land	Land to grow food and fibre for humans, animal fodder, oil palms, alcohol, etc.	Cut down on food intake
Grazing land	Land for animal husbandry for the production of meat, milk, skins and wool	Cut down on meat and dairy consumption
Forest	Land to supply timber, pulp for paper and fuel wood	Reduce paper usage – recycle
Built-up land	Land for infrastructure – houses, transport, reservoirs and power production	Have a smaller house
Water	Amount of water for all types of consumption – domestic, industrial, recreational and food production	Reduce fish consumption Shower rather than bathe Turn off tap while cleaning teeth

▲ **Figure 2.5** Aspects of ecological footprints and possible solutions

These are just a few suggestions for your questionnaire; www.Footprintnetwork.org has some other suggestions on how to reduce the ecological footprint (http://www.footprintnetwork.org/en/index.php/GFN/page/footprint_calculator_frequently_asked_questions/#gen6).

You can focus on one aspect of the ecological footprint or you can select one from each area. You can use either style of questionnaire (see Section 3c) for this investigation. Always make sure your questions are relevant to the investigation.

EXAMPLE: Bipolar questionnaire

In order to reduce your personal footprint would you?	−2 No way	−1 Unlikely	+1 Maybe	+2 Definitely
Install solar panels				
Cut down your food intake				
Stop eating meat				
Use recycled paper				
Live in a small house				
Shower instead of taking a bath				

EXAMPLE: Regular questionnaire

Which of the following measures would you be willing to take in order to reduce your water footprint?

Replace your bath tub with a shower	
Turn the tap off while you are cleaning your teeth	
Wash your car less often	
Buy a water-efficient washing machine	
Pay more for your water	

This could be followed by a number of questions linked to other aspects of the ecological footprint. Always make sure your questions are relevant to the investigation.

Suggested research question:

What is the relationship between ecological footprint and income (US$/month) or education level (years in school)?

Answer using data collected using questionnaires.

1. Design questionnaire.
2. Select sampling method.
3. Print at least 30 questionnaires (preferably more).
4. Select suitable locations to collect the data (think carefully).
5. Conduct the questionnaires.
6. Collate the questionnaires (if necessary).
7. Process and present the data.

Tip

Ecological footprint (primary data)

- Develop a sharply focused research question that is clearly linked to a specific environmental issue.
- Ecological footprints have both global and local significance but with a limited word count you may have to focus on one of them.
- Questionnaires are very time-consuming to develop so explain clearly how you designed it and why you used it.
- Make sure you pick an appropriate sampling strategy and justify it.
- Give a clear risk assessment and discuss ethical issues.

This investigation involves the collection of primary data about the country in which you live. You can design a questionnaire in order to investigate the ecological footprint and you can link it to income or education level. The question about income/education must be handled carefully.

You can focus on one aspect of the ecological footprint (food, travel, water, goods, etc.) or you can ask one question on a number of aspects. Remember the rules for questionnaires.

For the questions about the ecological footprint you can use some of the footprint calculators as a source of inspiration for your questions (remember to acknowledge whatever sources you use).

There are many footprint calculators; these are just a few suggestions:

* http://footprint.wwf.org.uk/ – this asks about food, travel, home
* http://www.footprintnetwork.org/en/index.php/GFN/page/ calculators/ – this asks about food, goods, shelter, mobility
* http://www.conservation.org/act/live_green/Pages/ecofootprint.aspx – this asks about recycling, pollution, shopping, food, energy efficiency, transport.

To make analysis and data presentation easier it is best to 'weight' the responses.

In this investigation you can give the 'best' response the lowest value or the 'worst' response the lowest value. So long as you are consistent it does not matter which option you go with.

SAMPLE QUESTIONS

It is your job to design the questionnaire but this is a sample.

1. What level of educational qualification do you have?

Completed school with no formal qualifications	
Completed secondary school at 16 with IGCSEs/GCSE (Use the most relevant one to your area)	
Completed secondary school at 18 with IB diploma/'A' levels (Use the most relevant one to your area)	
Went to university and have an undergraduate degree – BA, BSc	
Went to university and have a higher degree	

2. How often do you eat meat?

Never	
Once or twice a week	
Daily	
Twice a day	

You should have a maximum of 5–7 questions in the questionnaire but remember to always make sure your questions are relevant to the investigation.

SAMPLE QUESTION WITH WEIGHTING: the weighting should not be on the questionnaire you use to conduct the survey.

Key term

Weighting assigns a numerical value to each response. This makes graphical representation possible and thus analysis easier.

3. How often do you eat meat?

Question	Response	Weight
Never		1
Once or twice a week		2
Daily		3
Twice a day		4

This weighting is working with the idea that lower is better. So the best ecological footprint will be the lowest one.

- Once you have designed your questionnaire you need to decide on a sampling strategy and go out and collect your data.
- Then the data must be collated and summarized to facilitate clear data presentation.

How often do you eat meat?										
Level of education		Completed school with no formal qualifications		Completed secondary school at 16 with IGCSEs /GCSEs		Completed secondary school at 18 with IB diploma/'A' levels		Went to university and have an undergraduate degree – BA, BSc		Went to university and have a higher degree
	Weight (w)	Responses (r)	w*r	Responses (r)	w*r	Responses (r)	w*r	Responses (r)	w*r	Responses (r)	w*r
Never	1	1	1	4	4	3	3	3	3	0	0
Once or twice a week	2	2	4	3	6	8	8	3	6	1	2
Daily	3	3	9	4	12	2	6	1	3	1	3
Twice a day	4	5	20	5	20	3	12	2	8	1	4
Total response		11	34	16	42	12	29	9	20	3	9
Mean EFP		3.09		2.63		2.42		2.22		3.00	

▲ **Figure 2.6** Sample of collated processed data

Sample calculation

Completed school with no formal qualifications

Total response $= 1+2+3+5 = 11$

Mean EFP $= $ w*r/ Total responses

$\quad\quad\quad = 34/11$

$\quad\quad\quad = 3.09$

The complete set of collated processed data can now be presented in meaningful graphical form.

This could be followed by a bipolar questionnaire similar to the one earlier (page 20). This could then inform your discussions about solutions.

4. Investigate pollution

Topics 1 and 4

Suggested research question:

What are the sources and impacts of stream pollution and some possible solutions to the problem?

The example here is directed at locating a source of pollution in a stream system in the UK – the methods are adaptable to other types of pollution and other situations.

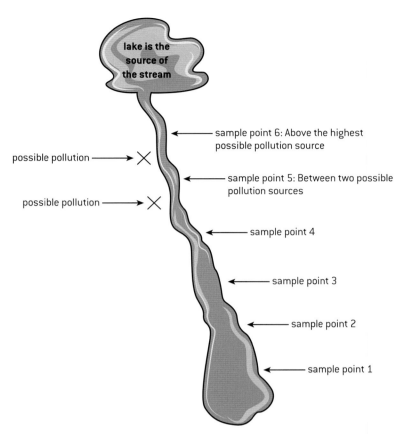

▲ **Figure 2.7** Map of a stream system and possible pollution sources

There are a number of ways to identify the presence of pollutants in a freshwater ecosystem

1. electronic probes

2. biotic index

Method

1. As you get into the river to take samples or conduct kick sampling you will cause material and invertebrates to be washed downstream; therefore you should start downstream and work upstream.

2. Sample points in river studies will often be determined by accessibility – try and sample as regularly as possible.

3. At each sample point

   a. if you are using pollution probes take all measurements before you get into the stream and disturb the sediments too much

   b. take a water sample to take back to the lab for testing – ensure that the container is clean and can be tightly sealed, and label the container with the sample point

   c. get into the river and do a kick sample (see page 64).

4. If sample point 5 is clear of pollution the source of pollution must be upstream of point 5. If point 5 is still showing pollution then you must take a sample at point 6 to make sure that the water upstream of sample point 5 is actually clean. That confirms which site is the source of pollution.

Data processing and presentation

The data you have collected is spatial – therefore your presentation should have a spatial element.

Tip

- Make sure you describe the methods of data collection clearly and justify sampling strategies.

- Pollution studies have significant associated risks so make sure you outline the appropriate safety precautions taken (e.g. when putting your hands in the water, you should wear gloves).

- You are dealing with live animals – discuss the ethical issues involved.

- The environmental issue is obvious so do not forget to discuss and evaluate the solution in detail.

For each sample point you can calculate the Simpson diversity index – these could then be shown on a bar graph or line graph.

Kite charts are good for showing the data spatially or you could use a bar graph for each site and place them on a copy of the map.

Then you can discuss the solutions to the problem of pollution.

Suggested research question:

What impact does intensive agriculture have on a local aquatic and terrestrial ecosystem?

Intensive farming is high-input farming. High yields are obtained by use of machinery, pesticides and fertilizer, all of which can potentially pollute local ecosystems.

1. Machinery uses energy in the form of diesel fuel causing air pollution – CO_2 and other exhaust gases/particulates.

2. Pesticides spread by wind during crop spraying may enter adjacent terrestrial ecosystems or end up in groundwater and in streams resulting in decreased biodiversity in both aquatic and terrestrial ecosystems.

3. Fertilizers usually contain nitrate, ammonium and phosphate ions. If too much fertilizer ends up in groundwater and streams this can lead to eutrophication.

These impacts can be easily assessed.

In aquatic ecosystems

You can use similar techniques to the previous example, except with this type of pollution you only need to take measurements upstream and downstream of the field.

You could run a range of tests on abiotic factors such as:

1. testing for fertilizers (nitrate, ammonium and phosphate ions) using test kits

2. temperature

3. turbidity, using a Secchi disc (see page 58)

4. salinity

5. dissolved oxygen.

And any others that you think may be affected by the intensive agricultural practices. Techniques for measuring pesticides do exist but are usually too complicated for school use.

The impact of differences can then be seen by assessing biotic factors including biodiversity using the Simpson diversity index. This is an indirect measure of the impact of both pesticides and fertilizers. (Use the kick sampling method, see page 64)

In terrestrial ecosystems

Intensive agriculture also impacts surrounding terrestrial ecosystems. To test for such impacts you could run transect lines (minimum of three) from the edge of the agricultural field into the natural area. If the agriculture is having an impact on the ecosystems there should be changes along the transect line.

NOTE: Your results will be better if you determine the dominant wind direction so you know which way the chemicals will be blowing.

Abiotic factors to look at:

1. soil pH, salinity, nutrient content, infiltration rates, moisture content and organic content

2. temperature.

Biotic factors to measure:

1. plant species abundance

   a. density

   b. frequency

   c. percentage coverage

   d. species diversity.

2. animal species present.

Suggested research question:

What is the relationship between water footprint and income (US$/month) or education level (years in school)?

Topics 4 and 8

This is very similar to the investigation about ecological footprints (page 18). The technique can be followed in much the same way, **except** that you are only asking questions about the water footprint. You can also do this as a two-part investigation. The same warnings apply!

There are a number of websites that may give you inspiration for questions to ask on your questionnaire (remember to credit sources).

SAMPLE QUESTION

It is your job to design the questionnaire but this is a sample.

What is your monthly income?

- Use local currency.
- Base your categories on the country you are in – the lowest should reflect the minimum wage.

Below 150	
150–500	
501–1000	
1001–1500	
More than 1500	

How long do you stay in the shower?

Less than 5 minutes	
5–7 minutes	
8–10 minutes	
More than 10 minutes	

You should have a maximum of 5–7 questions in your questionnaire but remember to always make sure your questions are relevant to the investigation.

SAMPLE QUESTION WITH WEIGHTING: the weighting should not be on the questionnaire you use to conduct the survey.

How long do you stay in the shower?

Question	Response	Weight
Less than 5 minutes		1
5–7 minutes		2
8–10 minutes		3
More than 10 minutes		4

This weighting is working with the idea that lower is better. So the best water footprint will be the lowest one.

- Once you have designed your questionnaire you need to decide on a sampling strategy and go out and collect your data.
- Then the data must be collated and summarized to facilitate clear data presentation.

		How long do you spend in the shower?									
Monthly income ($)		Below 150		150–500		501–1000		1001–1500		More than 1500	
	Weight (w)	Responses (r)	w*r	Responses (r)	w*r	Responses (r)	w*r	Responses (r)	w*r	Responses (r)	w*r
Less than 5 minutes	1	5	5	4	4	2	2	1	1	0	0
5–7 minutes	2	3	6	4	8	3	6	2	4	1	2
8–10 minutes	3	2	6	3	9	4	12	5	15	2	6
More than 10 minutes	4	1	4	2	8	4	16	5	20	6	24
TOTAL		11	21	13	29	13	36	13	40	9	32
Mean Water FP		1.91		2.23		2.27		3.08		3.56	

▲ **Figure 2.8** Sample of collated processed data

Sample calculation: Income below $150/month

Total response $= 1+2+3+5 = 11$

Mean EFP $= $ w*r/ Total responses

$= 34/11$

$= 3.09$

The complete set of collated processed data can now be presented in meaningful graphical form.

You could add an 'attitudes to solving the problem' survey.

The questions in this can be different from the ones in your original survey or you can keep the same emphasis.

In order to reduce your personal water footprint would you?	−2 No way	−1 Unlikely	+1 Maybe	+2 Definitely
Shower instead of taking a bath.				
Take a shorter shower.				
Shower less often in a week.				
Wash your car less often.				
Water your garden less often.				
Change the plants in your gardens to ones that demand less water.				
Get rid of your swimming pool.				

Suggested research question:

What are the changes in tropospheric ozone away from a source of pollution (e.g. a road) and what are the attitudes towards solving the problem?

Topic 6

This is a two-part investigation:

- tropospheric ozone concentrations
- attitudes towards solving the problem.

Part 1: Investigating changes in tropospheric ozone away from a source of pollution (e.g. a road)

Tropospheric ozone is mainly the result of the combustion of fossil fuels (see sub-topic 6.3) and it is possible to test for its presence using:

- shop-bought test papers specifically designed for the purpose
- Schönbein Paper which you can make yourself (see below).

Method

1. Select your study area:
   - line transect away from a road, or
   - line transect up a hill.
2. Place the testing strips every X metres (interval will depend on the area to be covered).
3. Make sure you place three testing strips at every interval and that you have a minimum of five intervals along your transect line.
4. Repeat the procedure every day for 10 days.
5. You can then process and present the data using the most appropriate method.

You can find out how to make testing strips at:

http://teachertech.rice.edu/Participants/lee/tropo.html

Tip

- The context is very straightforward – that is dangerous so check the criteria and make sure you have everything you need: a sharply focused research question based on an environmental issue (global or local).

- If you use shop-bought test papers you do not need to explain how they were bought! But you must explain clearly how they are used and justify the sampling strategy used.

- If you make your own Schönbein Paper using the method described at the internet link given on the left you must explain how the paper was made and cite the source.

- When using chemicals it is essential to discuss risk.

- Part 2 of the investigation leads on to a solution but you must remember to discuss and evaluate it.

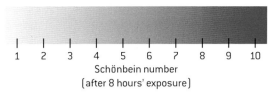

| | | | | | | | | | |
|1|2|3|4|5|6|7|8|9|10|

Schönbein number
(after 8 hours' exposure)

▲ **Figure 2.9** Schönbein Paper colour chart.

Part 2: Investigating people's attitudes to the solutions to tropospheric ozone

Remember this is about the problems of tropospheric ozone so solutions will be linked to reducing the combustion of fossil fuels. One way to assess people's attitudes to the problem and whether or not they can help is a questionnaire similar to the ones in Topic 1 and Topic 4.

In order to help reduce tropospheric ozone would you?	−2 No way	−1 Unlikely	+1 Maybe	+2 Definitely
Stop using your car and use public transport.				
Turn the lights off when you leave the room.				
Stop using air-conditioners/ heating in your home.				
Grow your own food or buy only local produce.				
Use eco-friendly products in your home.				
Buy energy-efficient products.				

The list of questions you could ask here is extensive – the focus is on reducing the combustion of fossil fuels.

5. Investigate carbon emissions

As with ecological footprint and the water footprint this can be approached in a number of ways:

- secondary data collection followed by a survey on environmental attitudes to the problems and the solutions
- primary data collection in the form of a questionnaire.

Suggested research question:

What is the relationship between carbon footprint and wealth/ education level/level of development of a country?

Part 1: Secondary data

See Topic 1 (page 16) for the secondary data collection method.

Possible websites for data on carbon emissions:

- http://www.nationmaster.com/
  - go to categories
  - select environment
  - select CO_2 emissions.
- http://www.gapminder.org
  - go to data tab
  - go to search – type Carbon
  - there are three different carbon emissions statistics.
- http://www.indexmundi.com/
  - click on environment
  - go to scroll down to CO_2 emissions
  - there is a wide range of data on CO_2 emissions.

See page 18 for possible websites giving wealth education and level of development.

Part 2: Primary data collection

See Topic 1 (page 19) for the primary data collection method for attitudes. You can use the questionnaire in this section or you can use the bipolar format shown below.

Possible questions for the bipolar or regular questionnaire on carbon emissions could ask people how prepared they would be to:

1. Install solar panels.
2. Pay higher prices for goods due to carbon taxes imposed on industry.
3. Walk or cycle instead of drive.
4. Take public transport instead of drive.
5. Car-pool for work or school.
6. Change driving style to reduce fuel consumption.
7. Drive fuel-efficient cars.
8. Check tyre inflation for increased fuel efficiency.
9. Fly less frequently.
10. Insulate their home to reduce use of electricity.

There are many more questions that could be asked in relation to this topic. Search the internet for 'Reducing carbon emissions'.

Bipolar questionnaire layout:

In order to reduce your carbon emissions would you:	−2 No way	−1 Unlikely	+1 Maybe	+2 Definitely
Install solar panels.				

Regular questionnaire:

Which of the following measures would you be willing to take in order to reduce your carbon emissions?

The respondent ticks all the measures they would be willing to take to reduce their carbon footprint.

Take public transport instead of drive	
Change your driving style to reduce fuel consumption	

Carbon footprint (primary questionnaire data)

The method for this is similar to the ecological footprint questionnaire (page 18).

Again there are plenty of footprint calculators on the internet and these can give you hints for possible questions to ask in your questionnaire. Suggested websites:

- http://cotap.org/carbon-footprint-calculator/
- http://www.carbonfootprint.com/calculator.aspx
- http://www.nature.org/greenliving/carboncalculator/#.

Once you have found the questions that suit your study you can design your questionnaire. It is useful to have an independent variable such as years in education/level of education or wealth. Once you have the results of this initial questionnaire you can do a second questionnaire about attitudes and you can use any of the surveys discussed in Section 3c.

For either of these investigations into carbon emissions you could add an EVS component to look at people's value systems linked to carbon emissions.

Have the table below on a separate sheet for respondents to view with the photos.

I am technocentric and I believe whatever problems we cause, we can solve them.	I am ecocentric and I believe we need the Earth more than it needs us.
<ul><li>We are the Earth's most important species, we are in charge.</li><li>There will always be more resources to exploit.</li><li>We will control and manage these resources and be successful.</li><li>We can solve any pollution problem that we cause.</li><li>Economic growth is a good thing and we can always keep the economy growing.</li></ul>	<ul><li>The Earth is here for all species.</li><li>Resources are limited.</li><li>We should manage growth so that only beneficial forms occur.</li><li>We must work with the Earth, not against it.</li></ul>

You do not have to use the same pictures as given here – there are plenty of others. You could also use one of the other bipolar analyses in this book.

Bigger and meaner hurricane surges expected in the future due to climate change		Tax on emissions is twice as burdening on the poor than on the rich	
	I am technocentric on this issue		I am technocentric on this issue
	I am ecocentric on this issue		I am ecocentric on this issue

The past decade saw unprecedented warming in the deep ocean		The 2014 headline: 'Global CO_2 level reaches 400 ppm for first time in human existence.'	
	I am technocentric on this issue		I am technocentric on this issue
	I am ecocentric on this issue		I am ecocentric on this issue

Global warming will lead to less snowfall in the world		Global warming will open shipping routes directly through the North Pole by 2050	
	I am technocentric on this issue		I am technocentric on this issue
	I am ecocentric on this issue		I am ecocentric on this issue

Lizards facing mass extinction due to global warming		Even if all emissions stop today, Earth will continue to warm for centuries	
	I am technocentric on this issue		I am technocentric on this issue
	I am ecocentric on this issue		I am ecocentric on this issue

IA examples

In this section are two examples of a complete IA. One is a weak example, the other is a stronger one. Each has comments and marks awarded based on IA criteria and markbands.

Please note that the marking and comments are those of a highly experienced ESS teacher and moderator but are not 'official' IB marks awarded.

Ecological footprint in Jordan – student A

Context

This report will look at the ecological footprint (EF) of Jordan using the following research question:

What is the ecological footprint of Jordan?

According to the WWF the ecological footprint is *"A measure of the impact humans have on the environment is called an ecological footprint. A country's ecological footprint is the sum of all the cropland, grazing land, forest and fishing grounds required to produce the food, fibre and timber it consumes, to absorb the wastes emitted when it uses energy and to provide space for infrastructure."*[1]

It would appear that our global EF is 1.5 – that means that we need 1.5 planets to supply everything we need and absorb all our wastes. This is a big problem because we don't have one and a half planets – we only have one and we are using too many of its resources and we will run out and destroy things. Not everyone has the same EF and there are big differences between countries. If you look at figure 1 you will see that richer countries have bigger EFs.

Figure 1 Ecological Footprint and human development[2]

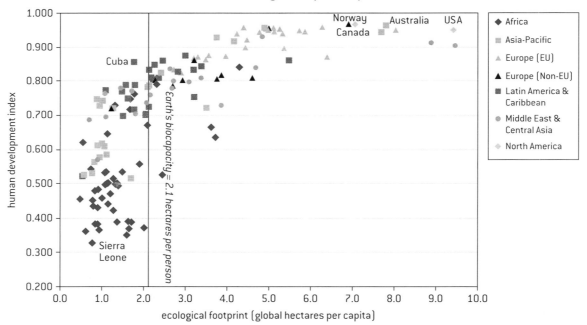

Data sourced from: Global footprint network 2008 report (2005 data). UN human development index 2007/08

According to global footprint network EFs have different aspects that are shown in figure 2. In the richer countries like the USA and Australia the people use more

---

[1] http://www.wwf.org.au/our_work/people_and_the_environment/human_footprint/ecological_footprint/

[2] https://en.wikipedia.org/wiki/Ecological_footprint#/media/File:Human_welfare_and_ecological_footprint.jpg

The research question is stated but lacks focus.

It is not good to use such large sections from the internet even if it is properly referenced.

This paragraph outlines the environmental issue but there is no detail.

- Energy – they have cars and lots of gadgets
- Settlement – they have lots of cities
- Timber and paper - they cut down forests for things
- Food and fibre – they eat more food and grow more crops
- Seafood – they eat lots of fish too.

> List of connections between the environmental issues and the research question.

In Jordan there is a big difference between the rich and the poor. So the rich people use a lot of resources and the poor don't use as much. So the issue here is that the EF is too big and the rich are using too many resources.

Figure 2 Ecological Footprint and human development[3]

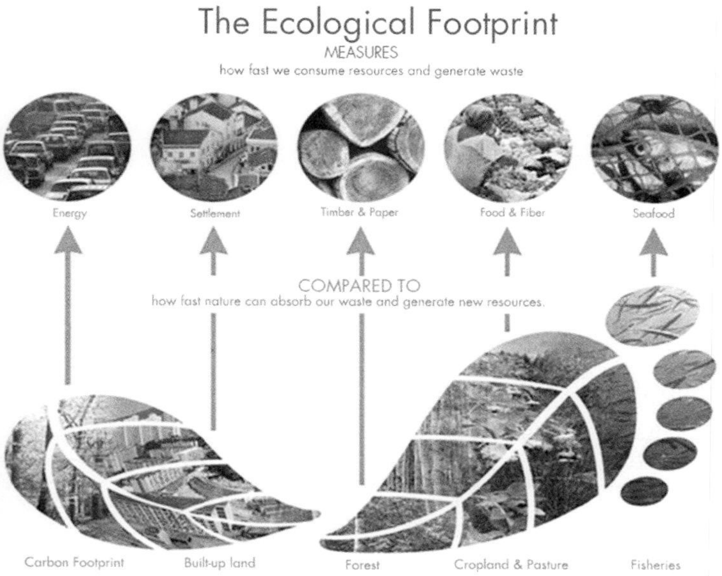

The Ecological Footprint
MEASURES
how fast we consume resources and generate waste

Energy   Settlement   Timber & Paper   Food & Fiber   Seafood

COMPARED TO
how fast nature can absorb our waste and generate new resources.

Carbon Footprint   Built-up land   Forest   Cropland & Pasture   Fisheries

---

Context: weak

- The research question is far too broad.
- The environmental issue is outlined but there is no discussion and it lacks depth.
- Connections are made between the research question and the environmental issue but they are listed and not explained.

SCORE 2 / 6

Planning

I am not going to measure the whole EF for Jordan but I will look at transport and compare it to how much people earn.

> 'I' is used throughout the report – inappropriate in an academic report.

Method:

1. Compile a questionnaire. To ask about how much people earn and what sort of transport they use.

> The method is appropriate and allows for the collection of plenty of data.

2. I have set 6 questions on transportation such as what kind of vehicle do you drive, how many hours do you spend on bus each week … I used the http://footprint.wwf.org.uk/ as a basis for most of the questions in my questionnaire. This is on the next page. Each response is weighted so I can later on calculate the EFP impact.

3. I will use systematic sampling to carry out the questionnaire in various places in Amman.

> Sampling strategy is stated but not explained.

[3] http://www.footprintnetwork.org/images/uploads/basics-overview-510.jpg

Questionnaire:

1. What range of income do you have per month in JOD? Please tick your answer.

Under 500	
501–750	
751–950	
951–1200	
1201+	

2. Which of these best describes the vehicle you use most?

Don't own a vehicle	
Motorbike	
Small petrol car	
Medium petrol	
Large petrol	
Small diesel	
Medium diesel	
Large diesel	

3. How many hours a week do you spend in cars or on motorbikes for personal use including commuting?

Under 2 hours	
2 to 5 hours	
5 to 15 hours	
15 to 25 hours	
Over 25 hours	

4. How long do you spend on the bus for personal purposes each week?

Under 1 hours	
2 to 5 hours	
5 to 7 hours	
7 to 10 hours	
Over 10 hours	

. How often do you travel by plane?

Never	
Weekly	
Monthly	
Less than 3 times a year	
3–5 times a year	

. Do you ever carpool?

Never	
Sometimes	
Most of the time	
Always	

Result analysis and conclusion

Figure 3: to show collated questionnaire results

1. Independent Variable – Income in JOD		Weighting	U 250	250–500	501–750	751–950	951–1200	1201 +
Total number of respondents			1	23	7	7	5	3
2. Which of these best describes the vehicle you use most?	don't own a vehicle	1	1	11	1	0	1	1
	small petrol car	2	0	8	2	3	2	2
	small diesel car	3	0	0	0	0	0	0
	medium petrol	4	0	2	4	0	0	0
	medium diesel	5	0	0	0	3	0	0
	large petrol	6	0	0	0	1	2	0
	large diesel	7	0	1	0	0	0	0
	motorbike	8	0	2	0	0	0	0
3. How many hours a week do you spend in cars or on motorbikes for personal use including commuting?	under 2 hours	1	0	1	1	0	0	0
	2 to 5 hours	2	0	2	1	0	1	1
	5 to 15 hours	3	0	8	2	4	1	0
	15 to 25 hours	4	0	2	3	2	1	1
	over 25 hours	5	0	1	0	1	2	0
	never	6	1	10	0	0	0	1
4. How long do you spend on the bus for personal purposes each week?	under 1 hours	1	0	6	2	3	1	0
	2 to 5 hours	2	0	7	0	0	0	2
	5 to 7 hours	3	0	3	1	2	1	0
	7 to 10 hours	4	1	2	0	0	0	0
	over 10 hours	5	0	0	0	0	0	0
	never	6	0	5	4	2	3	1
5. How often do you travel by plane?	never	1	1	16	5	4	1	0
	weekly	5	0	0	0	0	0	0
	monthly	4	0	0	0	0	0	1
	less than 3 times a year	2	0	6	2	2	2	1
	3 to 5 times a year	3	0	2	0	0	2	1

6. Do you ever carpool?	Never	4	1	19	5	6	5	3
	sometimes	3	0	4	1	1	0	0
	most of the time	2	0	1	1	0	0	0
	always	1	0	0	0	0	0	0

Qualitative data: during the process of asking people to fill out our questionnaire the weather was hot therefore I think that people have rushed while answering.

> Quantitative and qualitative data is clearly recorded though the qualitative data is limited.

Figure 4: shows the data for mean EF score for each question with a total for income brackets

	Mean EFP Score				
Questions	Under 500	501–750	751–950	951–1200	1201
Which of these best describes the vehicle you use most	2.3	3.3	5.0	3.4	2.3
How many hours a week do you spend in cars or on motorbikes for personal use including commuting?	2.8	④	4.6	4.8	③
How long do you spend on the bus for personal purposes each week ?	2.7	1.7	3.2	1.8	2.3
How often do you travel by plane	1.5	1.3	1.1	2.2	③
Do you ever carpool	3.9	3.6	3.6	④	④
TOTAL	13.2	13.9	17.5	16.2	14.6

> This is a good summary of the data but there are no sample calculations to show how the figures were obtained. And the decimal places are inconsistent – circled examples should be '3.0' or '4.0'

Figure 5 graph to show relationship between total mean EF and income.

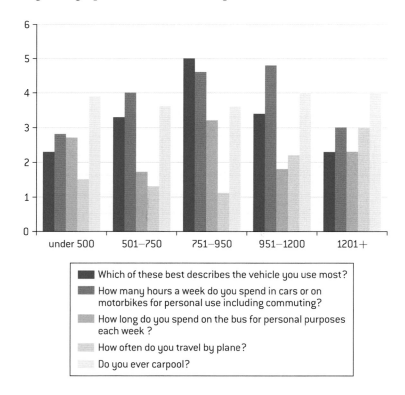

under 500   501–750   751–950   951–1200   1201+

■ Which of these best describes the vehicle you use most?
■ How many hours a week do you spend in cars or on motorbikes for personal use including commuting?
■ How long do you spend on the bus for personal purposes each week ?
■ How often do you travel by plane?
■ Do you ever carpool?

The graph is not that clear, too many bars make it hard to interpret. The axes are not labelled.

As for figure 5 which shows many trends about the data in which it tackles each question and income and its score on EF. In question one which describes your car the most; people under 500 JD as an income have scored a low EF which is 2.3, scoring such a low EF because people don't have the money to buy a car in the first place and others who own a car use a very small one to save money spent on petrol. As the higher income which is 751–950 with a EF of 5 this is because most of them have the ability to buy a car and maybe a huge one as well as many of them chose to buy a diesel one which contribute higher to the environment this is because it is cheaper for them to spend on for diesel. As for the highest income which is 1201 scoring lower than before with a 14.6 EF this is because they chose to buy a large petrol car which has lower EF making this income group in scoring lower EF than 751–950, this is because people with high income level experience a high level of education therefore people in this income bracket will have knowledge about recycling. As for the aspect of the car they owned the bracket with 751–950 had the largest car this is because of educational aspect as they are not educated on the environmental people with higher incomes tend to buy a suitable car for them not large than they need. As for the plane question people with higher income tend to use the plane for business issues and trading.

This describes results for question 1 and explains them but that is the only question tackled.

Result analysis and conclusion: Weak in places but okay on raw data.
- The graphs are overcomplicated and have no labels on the axes.
- Only one question is analysed so there are significant omissions.
- No conclusion stated.
  SCORE 1 / 6

Discussion and evaluation

The results show that it is not the richest people in Jordan that have the highest EF so they are not causing the bigger environmental issues. The middle-income people have bigger cars and spend more time driving. They may sometimes use the bus but they don't carpool. This has many environmental impacts.

This is a partial conclusion and it is related to the environmental issue.

Evaluation:

Limitations	Improvements
People may not be honest when answering	Stating that its 100% private such as putting the questionnaire in an envelope
Sampling is not accurate	Being more patient when sampling people, and places sometimes were crowded and others were not.
Weather	The weather was sunny and hot so people did not want to stand in the sun that much answering the questions

In the second row the weakness is very vague and the improvement does not make sense. In the third row the limitation is unclear and the 'improvement' is not an improvement.

Applications

If we know who has the highest EF in Amman we could use media campaigns to raise awareness of the problem. Maybe educate people about what they are doing. This would help reduce the EF if it worked but people may not care and so they will do nothing.

Bibliography

http://www.wwf.org.au/our_work/people_and_the_environment/human_footprint/ecological_footprint/

https://en.wikipedia.org/wiki/Ecological_footprint#/media/File:Human_welfare_and_ecological_footprint.jpg

http://www.footprintnetwork.org/en/index.php/GFN/page/footprint_basics_overview/

Ecological footprint in Jordan – student B

Context

This report will look at the ecological footprint (EF) of Jordan using the following research question:

What impact does wealth have on the transport aspect of the ecological footprint of Amman, Jordan?

The research question is relevant, clearly stated and focused.

Figure 1 Ecological Footprint and human development[3]

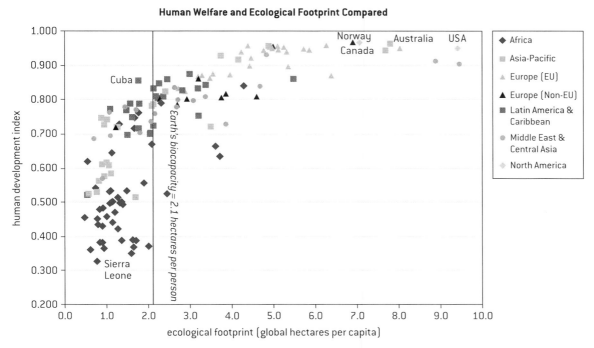

Human Welfare and Ecological Footprint Compared

Data sourced from: Global footprint network 2008 report (2005 data). UN human development index 2007/08

According to the IBO the EF is the area of land and water that is needed to provide a population with all the resources it needs and assimilate its wastes in a sustainable way. The EF can be expressed in global hectares per capita or as the number of planets we need to sustain our lifestyles. If our EF is bigger than the amount of land available or more than one planet then we are in trouble.

Figure 1 shows the Earth's biocapacity as 2.1 hectares per person. The biocapacity is the amount of biological materials the planet produces that are useful to humans. All the countries to the left of the line are ones that are living sustainably and they are nearly all in Africa, Asia-Pacific and Latin America & Caribbean. These are the LEDCs of the world. The MEDCs are mostly to the right of the line and are living unsustainably. This is obviously an issue because unsustainable living is ... well it is unsustainable, we cannot continue like this forever as resources will run out and pollution will kill the planet.

This paragraph starts to discuss the environmental issue

So why are MEDCs using so many resources? According to the Global Footprint Network EFs have different aspects. In the richer countries like the USA and Australia the people use more.

---

[3] https://en.wikipedia.org/wiki/Ecological_footprint#/media/File:Human_welfare_and_ecological_footprint.jpg

Energy:

MEDC societies have very high energy demands. Just about every home has an electricity supply and lots of things run on electricity – washing machines, lights, computers, kettles – the list goes on and on. Most people in MEDCs have at least one car per family and cars use petrol to run. Industry is well developed and a lot of the machinery runs on electricity. We live in a consumer society where stuff has built-in obsolescence, it is designed to run out and be thrown away so we buy new stuff.

In Amman there is very little public transport. People get around using their own cars or taxis. The cars are very old and cause a lot of pollution. Society in Jordan is changing and there is a growing middle class that wants to own their own car so the number of cars is increasing. The rich people have many cars per family and drivers will take them where they want to go then drive home and pick them up later so people cover a lot of miles. Many of the poorer people will drive diesel cars because it is cheaper than petrol, but they cause a lot of pollution.

> This paragraph links the environmental issue to the context.

Settlement:

According to internetgeography.net 74% of the people in MEDCs lived in cities in 1990. So cities take up a lot of space which means it is not being bioproductive. Not only that but cities use lots of resources, building materials for houses, roads and infrastructure, electricity for traffic lights and other road signs and for street lamps.

Timber and paper:

Forests are cut down for all sorts of things. Wood is used to make furniture, in buildings, fences, pencils etc, etc. We use a lot of paper too and that comes from some types of trees. We also cut down trees to make way for farming, building roads, mining minerals and enlarging our cities.

Food and fibre:

According to J H Lowry people in MEDCs eat 3340 calories and 90 grams of protein per day. About half the protein is animal protein. This is important in two ways – more calories means more food has to be grown to feed the people and more animal protein is very bad. It is better for people to eat lower down the food chain – the primary producers because there is less energy wasted. If you eat beef you have to use land to grow the cows' food and land for the cows to stay on. We also use a lot of land to grow crops that give us fibres – wool from sheep, linen from flax, and cotton.

Seafood:

As people become more health conscious they are changing their diet to include more fish and less meat. Increase demand for seafood means lots more fishing.

It would appear that our global EF is 1.5 – that means that we need 1.5 planets to supply everything we need and absorb all our wastes. This is a big problem because we don't have one and a half planets – we only have one and we are using too many of its resources and we will run out and destroy things. Not everyone has the same EF and there are big differences between countries. If you look at figure 1 you will see that richer countries have bigger EFs.

> This paragraph outlines the environmental issue but there is no detail.

Figure 1 showed that it is the MEDCs that have the largest EFs, this is a lot to do with their wealth – people can afford to use lots of energy, live in big houses, and eat lots of food and fish. So wealth seems like a good thing

> This paragraph links the environmental issue to the research question.

to look at when trying to work out what will affect the EF of people in Amman. Amman has about two million people from all sorts of income brackets so it will be easy to collect the data.

Hypothesis: The higher the income the higher the transport EF of people in Amman, Jordan.

> A hypothesis is not essential but it does help focus the investigation and it gives the student something to answer in the conclusion.

Context (CXT): strong

* The research question is relevant, coherent and focused.
* The environmental issue is relevant on a local and global scale and the background context for local and global levels is discussed
* Connections between the research question and the environmental issue are explained.
* The candidate may be penalized one mark as some of the explanation of the environmental issues is a little brief.

SCORE 5 / 6 or maybe 6 / 6

Planning

Method:

1. Compile a questionnaire. To ask about how much people earn and what sort of transport they use.

    a. Questionnaires are good because they give up-to-date data and can be designed to get the exact information you need, each questionnaire is quick to fill in and questions are the same for everyone.

    b. They do have the problem that people may not be honest because they are being asked about income; they are difficult to design.

> The method is appropriate and allows for the collection of plenty of data.

> Some advantages and disadvantages of questionnaires have been discussed.

2. There are 6 questions on transport and they were designed using http://footprint.wwf.org.uk/ as a basis. The questionnaire is on the next page. Each response is weighted to make it possible to calculate the EF impact. The lower the weight the lower the EF so things that reduce the EF should have the lowest number.

3. Systematic sampling is used to select the people to conduct the questionnaire in six different places in Amman. The locations must be in different areas that reflect different income groups. Systemmatic sampling is sometimes called every nth sampling because there is a set gap between sample points. In this study the person conducting the survey will stand in a strategic spot in the chosen location and interview every 10th person that passes them. 10 people will be interviewed at each location. Systemmatic sampling was chosen because it is easy to use. Random sampling is impossible because you have to know the entire population, give them a number and then select random numbers.

> Sampling strategy is stated and explained. Reasons are given for using systematic sampling but they are a little simplistic and nothing is mentioned about stratified sampling.

4. Risk assessment and ethical considerations.

    a. When addressing the public you must approach them politely and ask if they mind answering a few questions, do not be rude if they walk away.

    b. Point out that the income question is in brackets and that all information will be confidential.

    c. Always work in groups of three for safety.

    d. Take care when crossing any roads.

Sample of the questionnaire

1. What range of income do you have per month in JD?

Under 500	
501–750	
751–950	
951–1200	
1201+	

This question is not weighted because it is used to see how income affects the EF.

ALL WEIGHTS ARE GIVEN FROM BEST TO WORST FOR THE PLANET. 1 DOES THE LEAST HARM

2. Which of these best describes the vehicle you use most?

	Weight	
Don't own a vehicle	1	
Motorbike	2	
Small petrol car	3	
Medium petrol	4	
Large petrol	5	
Small diesel	6	
Medium diesel	7	
Large diesel	8	

The type of vehicles affects the size of the EF so they are in order by how many resources they use and so how much they impact the EF. The best thing is to not own a vehicle.

3. How many hours a week do you spend in cars or on motorbikes for personal use including commuting?

	Weight	
Under 2 hours	1	
2 to 5 hours	2	
5 to 15 hours	3	
15 to 25 hours	4	
Over 25 hours	5	

The more you drive the worse it is for the EF and the planet because you are probably using more resources.

4. How long do you spend on the bus for personal purposes each week?

	Weight	
Under 1 hours	1	
2 to 5 hours	2	
5 to 7 hours	3	
7 to 10 hours	4	
Over 10 hours	5	

This is better than owning a car but still the more you travel the more resources you use.

5. How often do you travel by plane?

	Weight	
Never	1	
Weekly	2	
Monthly	3	
Less than 3 times a year	4	
3 to 5 times a year	5	

Plane travel uses a lot of fuel so the less you travel by plane the better.

6. Do you ever carpool?

	Weight	
Never	4	
Sometimes	3	
Most of the time	2	
Always	1	

Carpooling means you share a car and there will be less resources used so it is best if you always carpool.

Planning (PI): Good

- The method is sound and will generate plenty of data and is explained clearly enough to be 'repeatable'. The weighting and questions are explained.
- Sampling strategy is explained and justified, though some elements are missing.
- Risk assessment and ethical considerations are discussed.

SCORE 5 / 6

Result analysis and conclusion

Figure 3: to show collated questionnaire results

Income in JOD		Weight	Under 500	501–750	751–950	951–1200	1201 +
Total number of respondents			25	7	7	5	3
Which of these best describes the vehicle you use most	don't own a vehicle	1	11	1	0	1	1
	small petrol car	2	9	2	3	2	2
	small diesel car	3	0	0	0	0	0
	medium petrol	4	2	4	0	0	0
	medium diesel	5	0	0	3	0	0
	large petrol	6	0	0	1	2	0
	large diesel	7	1	0	0	0	0
	motorbike	8	2	0	0	0	0
How many hours a week do you spend in cars or on motorbikes for personal use including commuting?	under 2 hours	1	1	1	0	0	0
	2 to 5 hours	2	3	1	0	1	1
	5 to 15 hours	3	8	2	4	1	0
	15 to 25 hours	4	2	3	2	1	1
	over 25 hours	5	1	0	1	2	0
	never	6	10	0	0	0	1

How long do you spend on the bus for personal purposes each week?	under 1 hour	1	7	2	3	1	0
	2 to 5 hours	2	8	0	0	0	2
	5 to 7 hours	3	3	1	2	1	0
	7 to 10 hours	4	2	0	0	0	0
	over 10 hours	5	0	0	0	0	0
	never	6	5	4	2	3	1
How often do you travel by plane?	never	1	16	5	4	1	0
	weekly	5	1	0	0	0	0
	monthly	4	0	0	0	0	1
	less than 3 times a year	2	6	2	3	2	1
	3 to 5 times a year	3	2	0	0	2	1
Do you ever carpool?	never	4	19	5	6	5	3
	sometimes	3	5	1	1	0	0
	most of the time	2	1	1	0	0	0
	always	1	0	0	0	0	0

Qualitative data:

It was a hot sunny day in Amman. We asked questions in different areas at different times but that is okay because the time of day does not affect these questions. At the first few locations people were in a rush to get to work so it was hard to get people to stop for us. Sometimes we went in to shops and asked the workers in the shops.

Figure 4: shows the data for mean EFP score for each question with a total for income brackets

> Quantitative and qualitative data are clearly recorded. There is not a lot of qualitative data but it is a suitable amount for this investigation.

	Mean EFP score				
Questions	Under 500	501–750	751–950	951–1200	1201+
Which of these best describes the vehicle you use most?	2.4	3.0	3.9	3.9	**1.7**
How many hours a week do you spend in cars or on motorbikes for personal use including commuting?	**4.2**	3.0	3.6	3.6	8.0
How long do you spend on the bus for personal purposes each week?	2.8	**4.1**	3.0	4.4	3.3
How often do you travel by plane?	1.6	1.3	1.4	2.2	3.0
Do you ever carpool?	**3.7**	**3.6**	**3.9**	**4.0**	**4.0**
TOTAL	14.7	15.0	15.8	18.1	20.0

Highlighted values are anomalies, discussed later.

Figure 5: Sample calculation for question 6. Do you ever carpool? Income bracket under 500 JOD

	Weight (w)	Responses (r)	w*r
never	4	19	76
sometimes	3	5	15
most of the time	2	1	2
always	1	0	0

Total EF score = 76 + 12 + 2 + 0 = 93

The mean EF is the total / number of respondents

93/25 = 3.7

> This is a good summary of the data sample; calculations are given to show how one of the sets of figures were obtained. Decimal places are consistent.

Figure 6: Stacked bar graph to show mean EF by income bracket and questions.

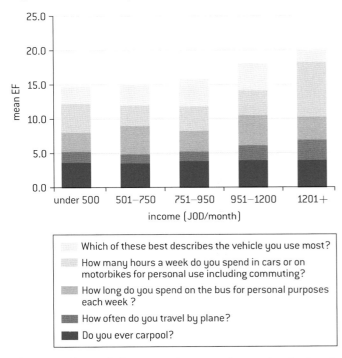

The overall trend shown in Figure 6 shows clearly that income has a significant impact on the transport EF of people in Amman, Jordan. People with the lowest income (under 500JOD/month) have the lowest mean EF of 14.7 then it steadily rises to 20.0 for people earning 1201+ / month. This is because generally people with less money cannot afford their own vehicle, they cannot afford to take the bus or the plane and they have no car to carpool with. There is a bit of an anomaly here though – they do have the second highest score for the question about how many hours they spend in a car / motorbike. This is probably for two reasons:

1. Amman has service taxis that are really cheap and shared by several people.

2. The bus service is very limited.

It is surprising that some of these people fly – maybe they did not really want to admit that they never flew anywhere. It is very important for people in Jordan to do the right things.

The richest people have the highest EF and this is because they have their own car which they drive everywhere in, they rarely use buses or carpool and they often travel by plane. The anomaly here is that they have a very low score for the question about the vehicle they drive. This is due to the fact that they can afford the most expensive cars with the best economy features. They have very high figure for time spent in their cars because they drive everywhere – even to the shop that is 100m away!

Other odd things are that people in the 501–750 JOD income/month bracket actually spend more time on the bus than the poorest people. It could be that they have good bus routes where they live so they don't use the service taxis.

The response to the carpooling was all very similar, most of the people never carpool, that is because people probably don't know about it. It would not be used for school because all schools have school buses.

With all this taken into account and the anomalies easy to explain the hypothesis can be accepted and it is true that the higher the income the higher the transport EF of people in Amman, Jordan.

Result analysis and conclusion (RAC): Good section with everything included.

- The collated data is given – this is appropriate for a questionnaire as it is not possible to show raw questionnaires. Qualitative data is given.

- The graph is clear and the axes are labelled. One graph is sufficient here. The stacked graph shows the total mean EF for each income bracket – that is necessary to be able to answer the research question. The breakdown by questions allows for a more detailed analysis to see where the biggest problems are.

- The results are analysed and patterns and anomalies identified.

- Trends are interpreted and explained in the context of the research question and a conclusion is drawn.

SCORE 6 / 6

Discussion and evaluation

So what does this all mean? The issue here is the size of our EF and it is clear that the higher the income the higher the EF in most aspects of transport. It is true that people with more money can afford to have the best cars with all the economy features that damage the environment less, but in Jordan that is the only thing that they do better. They scored well on the question about buses but that was because the question was badly done. The weighting should have been the other way around – the less time on buses the higher the score because it probably means you are using personal transport and that is not good. This probably means we need to start thinking about how the poorer people have lower EF and look at how they do that.

This paragraph evaluated the conclusion in the context of the environmental issue. It discusses who is causing the highest EF.

As stated this is not that realistic.

Evaluation:

Limitation	Impact	Improvement
People may not be answering honestly because they do not want to look bad.	Impact would not be too big because most people will do this so the results are the same for each income.	Use Survey Monkey or a postal survey for anonymous results. Though this may take a lot longer and not have as many responses. Could carry an envelope and let people complete the questionnaire and drop it in the envelope.

The number of people / income bracket was not even	May be quite serious because 1 odd result in 25 has less of an impact than 1 odd result in 3.	Stratified systematic sampling should be used to make sure all income groups were equally represented.
We did not really go to the very rich areas of Amman	The number of people in the highest income bracket was very low and this makes the results unreliable.	Make sure we visit the places where the richer people work or shop.
All data collection was between 9 am and 4 pm on weekdays.	Most people are at work so it was hard to get enough people.	Do some surveys in shopping areas or restaurants and cafes at the weekend.

Strengths

We did get a valid sample overall – more than 30 people and it was relatively easy to get the questionnaires done.

Modifications

1. The question about the buses was not very good and that should be changed. It may be better to have a question about who uses the bus and how often and not for how long. We should have done a pilot to see if the questionnaire worked well.

2. The questionnaire could also be done in other countries to compare actual LEDCs and MEDCs not just rich and poor.

3. It may have been better to ask questions about different aspects of the EF and not just transport, just one question on each aspect.

> Discussion and evaluation (DE): Good section
> - There is a conclusion with links to the environmental issue.
> - Weaknesses / limitations are discussed and realistic improvements suggested.
> - A couple of strengths are given and that may be all there is.
> - Modifications are discussed
> SCORE 5 / 6

Applications

There needs to be a two-part solution to this problem.

1. The government needs to be persuaded to improve the public transport network so people have an alternative to using their own car or taxis. AND

2. They could make it too expensive to drive in some parts of Amman – have a special road tax like London has.

This would mean that people would not have to have a car because they could use public transport. It would also benefit the poorer people as they could get around more easily. It may also make the centre of Amman less congested and that would reduce pollution and that would be good for the EF too.

This may work with some people but not the really rich ones because their car is a status symbol that they do not want to give up and the government could not charge enough to make them give it up.

WORD COUNT

2270 – with evaluation table

2069 – without the evaluation table

> That is 20 over the limit and moderators will count the words. You should declare your word count accurately. If you do not it is considered malpractice.

The word limit for this work is 1500–2250

You should not include the following in your word count

- Contents page (if you have one)
- Maps, charts, diagrams, annotations, illustrations and tables
- Equations, formulas and calculations
- Footnotes or endnotes
- Bibliography
- Appendices – BUT remember the moderator does not have to look at these so the report has to be valid without them.

Bibliography

IBO guide 2010 and 2017

Internetography

J H Lowry, World Population and Food Supply, (Edward Arnold, 1986) p. 24.

http://www.wwf.org.au/our_work/people_and_the_environment/human_footprint/ecological_footprint/

https://en.wikipedia.org/wiki/Ecological_footprint#/media/File:Human_welfare_and_ecological_footprint.jpg

http://www.footprintnetwork.org/en/index.php/GFN/page/footprint_basics_overview/

Practical scheme of work (PSOW)

Why a PSOW?

ESS has great potential for getting you out into the environment, doing fieldwork and collecting data. There are also lots of activities you can do in the classroom or laboratory. Your main assessed piece that counts towards the final exam is the internal assessment (IA) but you will also spend at least 20 hours on the PSOW.

> Practical work is any work you do in the classroom, the laboratory or outside in the field (or pond, beach, mountainside). 'In the field' is a term that scientists use to describe investigations and experiments done outside in an ecosystem.

The ESS guide says that you must carry out practical work.

This practical work will help you to develop the skills you will need for ESS.

Practical work should be totally integrated into the ESS course teaching.

The aims of the PSOW are to:

- reinforce what you have done in the theory lessons
- teach you useful laboratory and fieldwork data collection techniques that can be used in the IA
- help you find secondary databases – also for use in the IA
- teach you appropriate data presentation, analysis and evaluation techniques
- give you the opportunity to see the links between environmental and social systems.

During the PSOW you may do some or all of the following:

- short labs or longer projects
- computer simulations using databases for secondary data
- develop and use models
- conduct questionnaires, user trials and surveys
- data-analysis exercises
- fieldwork.

What is in this section?

In this section you will find:

- data collection techniques – these are general and can be widely applied
- suggested areas that the technique can be used in
- IA sections that can be practised.

It is divided into:

- fieldwork techniques – mostly sub-topic 2.5 of the ESS guide.
- questionnaires – could be used in most topics
- specific activities for each of the topics of the guide.

A word on ethical practice

- No experiments involving other people will be undertaken without their written consent and their understanding of the nature of the experiment.

- No experiment will be undertaken that inflicts pain on, or causes distress to, humans or live animals, in line with the animal experimentation policy, available on the OCC.

- No experiment or fieldwork will be undertaken that damages the environment.

Animal experimentation

All ecosystem investigations should follow the guidelines in the **IB animal experimentation policy**. This may be more stringent than your national standards so check it carefully before designing an experiment.

Consider if you could:

- replace the animal by using cells, plants or simulations

- refine the experiment to alleviate harm or distress

- reduce the number of animals involved.

The IB policy states that you may not carry out an animal experiment if it involves:

- pain, undue stress or damage to the health of the animal

- death of the animal

- drug intake or dietary change beyond those easily tolerated by the animal.

If humans are involved, you must also have their written permission and not carry out experiments that involve the possibility of transfer of blood-borne pathogens.

Academic honesty

All coursework – including work submitted for assessment – is to be authentic, based on your individual and original ideas with the ideas and work of others fully acknowledged. Assessment tasks that require teachers to provide you with guidance or that require you to work collaboratively must be completed in full compliance with the detailed guidelines provided by the IB for the relevant subject.

What the IB says

> The internal assessment task involves the completion of an individual investigation of an environmental systems and societies research question that has been designed and implemented by you. Your investigation is submitted as a written report.
>
> Note: Any investigation that is to be used for IA should be specifically designed by you to address the assessment criteria. You should therefore be provided with a copy of the assessment criteria when the requirements of the investigation are explained.
>
> If you undertake an extended essay in ESS, it must not be based on the same research question as the IA.

You will find it helpful to have:

- a list of all practical skills listed in the ESS guide
- a list of statements involving local examples or case studies.

Documents

Make sure you have copies of:

- the ESS guide
- the IB animal experimentation policy.

PSOW ideas

Here are some ideas for activities that could form part of a PSOW. Investigations in bold type have detailed information in the next section of this book.

Topics	Practical work ideas
Topic 1: Foundations of environmental systems and societies 1.1 Environmental value systems 1.2 Systems and models 1.3 Energy and equilibria 1.4 Sustainability 1.5 Humans and pollution	• **Set up an aquatic or terrestrial ecosystem in a bottle (remember to comply with the animal experimentation policy). http://www.bottlebiology.org/** • **Investigate feedback in computer simulation. http://www.goldridge08.com/flash/fc44/foodchain.swf** • Investigate a candle/pot plant/kettle/human as a system. • Investigate a local ecosystem (stream/ditch/pond/tree/bush/garden) as a system and measure inputs/outputs/flows/storages. • Identify environmental indicators for your school and measure their change over time, e.g. pollution levels of noise/dust in air/acid pollution.
Topic 2: Ecosystems and ecology 2.1 Species and populations 2.2 Communities and ecosystems 2.3 Flows of energy and matter 2.4 Biomes, zonation and succession 2.5 Investigating ecosystems	• Bottle biology as in Topic 1. • **Build a food chain for a local ecosystem.** • Compare brine shrimps kept in different conditions, e.g. light/dark, with pond weed/without pond weed, different temperatures. (Remember to reduce the risks to the shrimp.) • Investigate the ecological relationships in a given area, e.g. producers/consumers/decomposers/predators/prey/mutualism/parasitism. • Investigate the presence or absence of zonation patterns in local ecosystems. • Collect and identify organisms from the school/a local pond (always put them back where you found them). • Use identification keys to find out what organisms are present. Draw up a food web having researched their foods. • Investigate biomass of trophic levels in a local ecosystem (use secondary data to avoid killing animals). • Investigate differences in diversity for two or more different habitats. • Investigate factors affecting the rate of photosynthesis. • Investigate primary productivity in grassland or an aquatic plant (light and dark bottles). • Measure decomposition rates of materials under different conditions, e.g. paper/plastic bags/wood/organic waste/orange peel. • Investigate changes in productivity in different habitats of an ecosystem that you have visited.

	<ul><li>Investigate secondary productivity using a simulation or secondary data.</li><li>Carry out data analysis of sustainable yields.</li><li>Analyse data on population growth of different species.</li><li>Investigate the changes occurring along a transect line. Choose one of the following areas to investigate: (a) type of material, (b) level of pollution, (c) soil type, (d) temperature or (e) amount of moisture.</li><li>**Investigate the factors affecting the accuracy of estimates made using the Lincoln index (capture–mark–release–recapture).**</li><li>Use or plan use of Lincoln index to estimate size of a population, e.g. your school/ town.</li><li>Simulation of vegetation sampling techniques.</li><li>**Construct dichotomous keys.**</li><li>Evaluate Simpson's diversity index using a simulation.</li><li>Compare native forest with planted monoculture.</li><li>Investigate the human impact on a given area.</li><li>Use a sampling method to quantify the pattern and measure changes in an abiotic factor that is considered responsible for the gradient.</li><li>Design an investigation that will look at a limiting factor on plant, animal or fungi growth (for example, flour weevils, duckweed (*Lemna*), yeast).</li><li>Compare the biodiversity of two contrasting areas in an ecosystem.</li><li>Investigate the effect of altitude on species composition/change in one species.</li><li>Evaluate an EIA.</li><li>Investigate succession in an ecosystem of your choice.</li></ul>
Topic 3: Biodiversity and conservation 3.1 An introduction to biodiversity 3.2 Origins of biodiversity 3.3 Threats to biodiversity 3.4 Conservation of biodiversity	<ul><li>**Investigate global biodiversity.**</li><li>**Investigate the relationship between number of threatened species and the deforestation rates of a country.**</li><li>**Investigate the differences in species diversity in pools and riffles.**</li><li>**Evolution/natural selection simulations.**</li><li>Scatter 1 cm pieces of coloured straw (100 of each colour) on a small area of grass. Students collect as many as they can in 30 seconds and calculate the percentage they found of each colour. More camouflaged will be harder to find so more 'survive'.</li><li>Visit a local zoo and/or botanical garden and write case studies of endangered species there.</li><li>Visit a protected/conservation area.</li><li>Interview local environmental organizations.</li></ul>
Topic 4: Water and aquatic food production systems and societies 4.1 Introduction to water systems 4.2 Access to freshwater 4.3 Aquatic food production systems 4.4 Water pollution	<ul><li>Investigate the amount of renewable freshwater available for 30 selected countries.</li><li>**Investigate the total water withdrawal for 30 selected countries.**</li><li>**Investigate fishing rates in selected countries.**</li><li>**Investigate EVSs regarding water use.**</li><li>**Investigate people's attitudes and EVSs to problems caused by exploitation of the oceans' resources.**</li><li>**Measure the biochemical oxygen demand in a variety of water sources.**</li></ul>

	• **Investigate the impact of nitrates or detergents on the growth of pondweed (duckweed).**
	• **Investigate the impact of different surface material on the rate of infiltration to assess the possible impacts on flooding.**
	• Measure water use in your household over one day; pool class results.
	• Compare the Biological Oxygen Demand of water samples under different conditions.
	• Compare water quality of a variety of samples.
	• Investigate aquatic, terrestrial or air pollution in your city, town or college through the use of the presence or absence of indicator organisms (e.g. lichens, mayflies).
Topic 5: Soil systems and terrestrial food production systems and societies 5.1 Introduction to soil systems 5.2 Terrestrial food production systems and food choices 5.3 Soil degradation and conservation	• **Investigate differences in soil profiles locally.** • **Investigate soil erosion.** • **Investigate impact of salinization on plant growth or seed germination.** • **Investigate food consumption and/or production patterns.** • **Investigate your food consumption.** • Compare the characteristics of three types of soil. • Field trip to study soil conservation strategies. • Investigate the attitudes and methods of a local farming system to soil conservation. • Visit a farm and interview the farmer. • Compare meat consumption among different populations. • Plan an investigation into one of these aspects of soil: (a) compaction, (b) soil conditioners, (c) wind reduction techniques or (d) cultivation techniques. • Plan and carry out an investigation into the factors that affect sediment load in run-off.
Topic 6: Atmospheric systems and societies 6.1 Introduction to the atmosphere 6.2 Stratospheric ozone 6.3 Photochemical smog 6.4 Acid deposition	• **Investigate acid deposition in a range of countries.** • **Investigate air pollution using a biotic indicator (lichens).** • **Investigate the impact of albedo of different surfaces on the temperature above them.** • Investigate the effect of acid deposition on an aspect of plant growth. • Investigate the effect of ozone depletion on an aspect of plant growth. • Investigate the effect of pollution of an aspect of plant growth. • Investigate changes in particulate pollution with distance from source. • Use secondary data to compare the relationship between weather and air pollution. • Design an experiment to measure the effect of acid rain on either plants or building materials. • Design an experiment to look at the effects of ultraviolet radiation on plants or materials, e.g. wheat, rubber, plastic, different fabric T-shirts. • Place sticky traps (glass slides covered in a thin layer of Vaseline or sticky tape) around the school. Leave for 1, 2 or 4 weeks and examine under a dissecting microscope or with a hand lens, categorize and count findings and relate to positions of traps.

Topic 7: Climate change and energy production	•	**Investigate the impact of different gases on the temperature of a body of air.**
7.1 Energy choices and security	•	Investigate the global warming potential of gases.
	•	Investigate the albedo effect of various substrates.
7.2 Climate change – causes and impacts	•	Investigate perceptions of global warming.
7.3 Climate change – mitigation and adaptation	•	Visit an electricity generation site, e.g. a nuclear power station/fossil fuel fired power station/anaerobic digester.
	•	Calculate the carbon footprint of your household or your school.
Topic 8: Human systems and resource use	•	**Graphically represent changes in human population size over time.**
	•	**Draw a population pyramid in Excel.**
8.1 Human population dynamics	•	**Draw compound line graphs to show changes in birth rates, death rates and growth rates over time for a country of your choice.**
8.2 Resource use in society	•	**Investigate the relationship between level of development and demographics.**
8.3 Solid domestic waste	•	**Investigate environmental value systems (EVS) and resource exploitation.**
8.4 Human population carrying capacity	•	**Investigate environmental attitudes towards resource exploitation.**
	•	**Investigate solid domestic waste generated in school compared to your home.**
	•	**Investigate attitudes towards the intrinsic value of landscapes.**
	•	**Simulation on running your own country.**
	•	Investigate your ecological footprint and make a comparison with others.
	•	Tragedy of the commons simulation http://es.earthednet.org/fishing-game.
	•	Visit or get data from a local landfill site – data on proportions of paper/food/plastic/metal/dust/ash. Find the catchment area. Compare with other sites in another region or country. Evaluate landfill vs other methods of waste management.
	•	Visit a recycling centre/water treatment works works/incinerator/anaerobic digestor.
	•	Measure (weigh and sort) the waste generated by the student/home/school for one week.

Fieldwork

Fieldwork data collection

In this section, we will explore some data collection techniques you could use as part of your practical work. Topic 2 in particular lends itself to traditional environmental studies and a number of basic techniques underpin many of the possible investigations. Once you understand these various methods of data collection you can combine them to collect the relevant data for a wide range of investigations.

Quadrats

▲ **Figure 3.1** Samples of quadrats

How many quadrat samples, and of what size?

The size of the quadrat chosen is dependent on the size of the organisms being sampled.

Quadrat size	Quadrat area	Organism
10 × 10 cm	1 m²	Very small organisms: lichens on tree trunks or walls, or algae.
0.5 × 0.5 m	0.25 m²	Small plants: grasses, herbs, small shrubs. Slow moving or sessile animals: mussels, limpets.
1.0 × 1.0 m	1 m²	Medium size plants: large bushes.
5.0 × 5.0 m	25 m²	Mature trees.

▲ **Figure 3.2**

There is a balance to strike between increasing accuracy with increasing size and time available and the number of times a quadrat is placed. These will vary depending on the ecosystem, size of organisms and their distribution. There is a simple method which will help you determine the appropriate number of samples:

As you increase the number of samples, plot the number of species found. When this number is stable, you have found all species in the area, so in Figure 3.3, eight samples are enough.

If you increase the size of the quadrat (e.g. from side length 10 cm, 15 cm, 20 cm and so on) and plot the number of species found, when this number reaches a constant, that is the quadrat size to use.

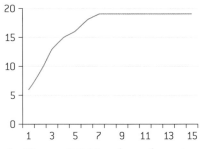

▲ **Figure 3.3** Number of species and quadrat size

How to place quadrats

Quadrats can be placed randomly, continuously or systematically (according to a pattern).

1. **Random quadrats** may be placed by throwing the quadrat over your shoulder, but we do not recommend this as it could be both dangerous and not random – you could decide where to throw.

   The conventional method is to use random number tables.

   - Map out your study area.
   - Draw a grid over the study area. (Figure 3.4)
   - Number each square.
   - Use a random number table to identify which squares you need to sample.

2. **Stratified random sampling** is used when there is an obvious difference within an area to be sampled and two or more sets of samples are taken.

   The area for study in Figure 3.5 has two distinctly different vegetation types and three separate areas to be studied. Samples need to be taken in each area.

   - Deal with each area separately.
   - Draw a grid for each area.
   - Number the squares in each area (they can be the same numbers or different).
   - Use a random number table to identify which squares you need to sample in each area.

3. **Continuous and systematic sampling** along a transect line.

   You might use this to look at changes in organisms as a result of changes along an environmental gradient, e.g. zonation along a slope, a rocky shore or grassland to woodland, or to measure the change in species composition with increasing distance from a source of pollution. Transects are quick and relatively simple to conduct.

There are two main types of transect that could be useful to you.

1. **Line transect**: consists of a string or measuring tape which is laid out in the direction of the environmental gradient and species touching the string or tape are recorded.

2. **Belt transect**: this is a strip of chosen width through the ecosystem. It is made by laying two parallel line transects, usually 0.5 or 1 metre apart, between which individuals are sampled.

Transect lines may be continuous or interrupted.

1. In a **continuous transect** (line or belt transect) the whole line or belt is sampled.

2. In an **interrupted transect** (line or belt) samples are taken at points along the line or belt. These points are usually taken at regular horizontal or vertical intervals. This is a form of systematic sampling. Quadrats are placed at intervals along the belt.

NOTE:
Many line transects (at least three, preferably five) need to be collected to obtain sufficient reliable data.

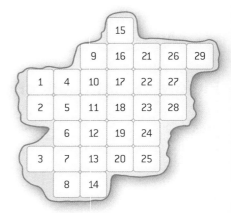

▲ **Figure 3.4** Numbering of random quadrats on a grid

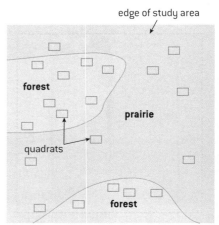

▲ **Figure 3.5** Random quadrat sampling

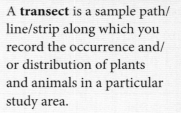

Key term

A **transect** is a sample path/ line/strip along which you record the occurrence and/ or distribution of plants and animals in a particular study area.

▲ **Figure 3.6** Laying out a transect line on a moor

Measuring abiotic components of the system

Ecosystems can be roughly divided into marine, freshwater and terrestrial ecosystems. Each of these ecosystem types has a different set of physical (abiotic) factors that you can measure.

Marine ecosystems

Abiotic factors: salinity, pH, temperature, dissolved oxygen, wave action to name just a few. Whichever abiotic factor you choose, remember they vary over space and time so be careful to control one of these factors. Many of these abiotic factors can be measured by using modern data loggers with interchangeable sensors or probes for pH, salinity, temperature or dissolved oxygen.

General method:

1. Select your abiotic variable and the appropriate probe.

2. Decide if you are going to measure the changes over time or space. Keep the one you are not changing constant.

    a. If you are measuring changes in temperature with depth try and take all readings at the same time (this can be difficult).

    b. If you are measuring changes in salinity through time take all measurements at the same place.

3. Each time you take your readings do at least five repeats so you can calculate a mean value and reduce errors.

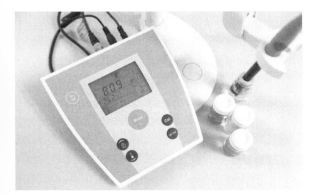

▲ **Figure 3.7** Example of a probe with a pH sensor

Dissolved oxygen can be measured using a Winkler titration. A series of chemicals is added to the water sample and dissolved oxygen in the water reacts with iodide ions to form a golden-brown precipitate. Acid is then added to release iodine which can be measured, and is proportional to the amount of dissolved oxygen, which can then be calculated. For a more detailed method check on the internet or ask your teacher.

Freshwater ecosystems

Abiotic factors: turbidity, flow velocity, pH, temperature, dissolved oxygen.

The methods for measuring pH, temperature and dissolved oxygen are the same as for marine ecosystems.

Turbidity

Turbidity can be measured with optical instruments or by using a Secchi disc. High turbidity = cloudy water; low turbidity = clear water.

A Secchi disc is a white or black-and-white disc attached to a graduated rope. The disc is heavy enough to ensure that the rope goes vertically down.

The procedure is:

1. Slowly lower the disc into the water until it disappears from view.
2. Read the depth from the graduated rope.
3. Slowly raise the disc until it is just visible again.
4. Read the depth from the graduated rope.
5. This should be repeated in the same spot 3–5 times so that the mean can be calculated and errors reduced. The mean reading is known as the Secchi depth.

For reliable results a standard procedure should be followed:

- Always stand or always sit in the boat.
- Always wear your glasses or always work without them.
- Always work on the shady side of the boat.

▲ **Figure 3.8** A Secchi disc

Flow velocity

This is the speed at which the water is moving and it determines which species can live in a certain area. Flow velocity varies with:

1. Time: melt water in the spring gives high flow rates, summer drought causes low flow rates.
2. Depth: surface water may flow more slowly than that in the middle of the water column.
3. Position in the river: inside bend has shallow slower moving water, outside bend has deeper fast moving water.

There are three basic methods for measuring flow velocity.

1. Flow meter: these are generally expensive and can be unreliable as mixing water with electricity has its problems.
2. Impellers: a simple mechanical device as shown in figure 3.9.
   a. The impeller is mounted on a graduated stick. It's base should be placed on the river bed. The height of the impeller can be adjusted and the velocity measured at different depths, BUT it can only be used in clear shallow water, as you must be able to see the impeller.
   b. The impeller is held at the end of the side arm and lowered into the water facing upstream.

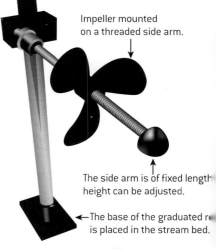

Impeller mounted on a threaded side arm.

The side arm is of fixed length height can be adjusted.

←The base of the graduated r is placed in the stream bed.

▲ **Figure 3.9** Impeller

   c. The impeller is released and the time it takes to travel the distance of the side arm is measured.

   d. Repeat 3–5 times for accurate results.

3. Floats: The easiest way to measure flow velocity is to measure the time a floating object takes to travel a certain distance. The floating object should preferably be partly submerged to reduce the effect of wind. Oranges and grapefruits can be used as floats but the water needs to be suitably deep for them. This method gives the surface flow velocity only. The average flow velocity of a river can be estimated from the surface flow velocity by dividing the surface velocity by 1.25.

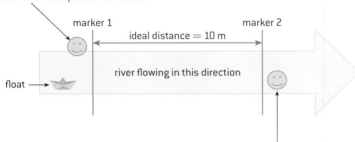

person 1 drops the float above the first marker and shouts 'start' as it passes the marker

marker 1

marker 2

ideal distance = 10 m

river flowing in this direction

float

person 2 starts the stopwatch on command from person 1 and stops it as the float passes marker 2 and catches the float

▲ **Figure 3.10** How to measure stream velocity

This should be repeated 3–5 times for accuracy.

WARNING: This methods gives seconds/metre NOT metres/second.

Terrestrial ecosystems

Abiotic factors: temperature, light intensity, wind speed, soil texture, slope, soil moisture, drainage and mineral content.

As with marine ecosystems, many of the abiotic variables of a terrestrial system can be measured using a data logger and an appropriate probe.

Air temperature

Temperature can be measured using simple liquid thermometers and min–max thermometers.

Wind speed

There are a variety of techniques used to measure wind speed:

- A revolving cup anemometer consists of three cups that rotate in the wind. The number of rotations per time period is counted and converted to a wind speed. Revolving cup anemometers can be mounted permanently or hand-held.

- A ventimeter is a calibrated tube over which the wind passes. This reduces the pressure in the tube, which makes a pointer move. It is easy to use and inexpensive.

- By observation of the effect of the wind on objects. The observations are then related to the Beaufort Scale (a scale of wind speed from 0 to 12).

Rainfall

Rainfall can be collected using a rain gauge. Some schools have an established weather station, in which case collecting rainfall data is easy. Many schools will not have a weather station but rain gauges are very easy to make and there are plenty of websites that can give you advice on how to make your own. Once you have made your rain gauge:

1. Place your rain gauge in a suitable spot in the study area – somewhere away from the influence of buildings, trees and other obstacles that may affect rainfall.

2. Check the rain gauge every 24 hours – at the same time every day. Pour rain into a graduated cylinder and record the daily amount of rainfall.

▲ **Figure 3.11** A rain gauge

Soil

Soil has a significant impact on plant growth and there are a variety of aspects of the soil that can be measured.

Soil texture (particle size)

Soil is made up of particles (gravel, sand, silt, clay) and the average size and distribution of them affect a soil's drainage and water-holding capacity.

Particle	How to measure
Gravel: very coarse, coarse and medium	Measure individually – simple, but time-consuming procedure.
Gravel: fine and very fine	Sieved through a series of sieves with different mesh sizes.
Sand: all sizes	
Silt and clay	Sedimentation or optical techniques. Sedimentation techniques are based on the fact that large particles sink faster than small particles. Optical techniques use light scattering by the particles (light scattering is what makes suspensions of soil particles in water look cloudy). Both sedimentation and light scattering can be done using automated instruments but they are expensive for secondary school use.

Soil moisture

This is the amount of water in the soil. It can be measured by drying soil samples.

1. Place a sample of the soil in a crucible.

2. Weigh it and record the weight.

3. Dry the sample.

Drying can be done in a conventional drying oven or a microwave oven.

In a conventional oven:

- Set the oven to 105 °C; hot enough to dry the soil but not so hot as to burn off organic matter.

- Leave for 24 hours and weigh the sample; repeat this until its mass becomes constant. This could take several days.

A minimum of 3–5 samples should be tested.

In a microwave oven:

- Place the sample in the microwave for 10 minutes.
- Weigh the sample, and return to the oven for 5 minutes – repeat until its mass becomes constant.

A minimum of 3–5 samples should be tested.

Soil organic content

The organic content of a soil is made up of plant and animal residues in various stages of decay and it has several functions.

- Supplies nutrients to the soil.
- Holds water (like a sponge).
- Helps reduce compaction and crusting.
- Increases infiltration.

Organic content can be determined by the loss on ignition (LOI) method.

1. Dry the sample as above and record the weight of the dry sample.
2. Heat the soil at high temperatures of 500 to 1000 °C for several hours.
3. Weigh the sample and repeat this until its mass becomes constant.

A minimum of 3–5 samples should be tested.

Soil mineral content and pH

There is a wide range of soil nutrients essential for a fertile soil. These are easy to measure through traditional soil testing kits or the ones available in many gardening centres.

Soil pH can also be measured using a soil testing kit or a pH probe.

Measuring biotic components of a system

Plant biomass

Measuring plant biomass is simple but destructive. Generally speaking it is best to take above-ground biomass as trying to get at parts below the ground such as roots can be very difficult.

For low vegetation/grasses:

1. Place a suitably sized quadrat (see figure 3.2).
2. Harvest all the above-ground vegetation in that area.
3. Wash it to remove any insects.
4. Dry it at about 60–70 °C until it reaches a constant weight. Water content can vary enormously so all the water should be removed and the mass given is dry weight.
5. For accurate results this should be repeated 3–5 times and a mean/unit area can be obtained.
6. The result can then be extrapolated to the total biomass of that species in the ecosystem.

For trees and bushes:

1. Select the tree or bush you which to test.
2. Harvest the leaves from 3–5 branches.
3. And repeat steps 3–6 in the above method.

Primary productivity

In **aquatic ecosystems** (both marine and freshwater ecosystems) the **light and dark bottle technique** can be used to measure both the gross and net productivity of aquatic plants (including phytoplankton). This is simple but has given us a good idea of the productivity of the oceans and of many lakes.

The productivity is usually calculated from the oxygen concentrations in the bottles. The procedure is:

1. Take two bottles filled with water from the ecosystem
   a. one of the bottles is made of clear glass
   b. the other is of dark glass or is covered to exclude light.
2. Measure the oxygen concentration of the water by chemical titration (Winkler method) or an oxygen probe, and record as mg oxygen per litre of water.
3. Place equal amounts of plants of the same species into each of the bottles.
4. Both bottles must be completely filled with water and capped. (No air should be present.)
5. Allow to stand and incubate for several hours.
6. Measure the oxygen levels in both bottles and compare with the original oxygen level of the water. The incubation can take place in the laboratory or outdoors in the ecosystem of investigation.

In the light bottle, photosynthesis and respiration have been occurring. In the dark, only respiration occurs.

In terrestrial ecosystems, you can do a similar experiment with square 'patches':

1. Select three equally sized patches with similar vegetation (e.g. grass).
2. The first patch (A) is harvested immediately and the biomass measured (see above).
3. The second patch (B) is covered with black plastic (no photosynthesis, just respiration).
4. The third patch (C) is just left as it is.
5. After a suitable time period (depends on the season), patches B and C are harvested and the biomass measured (as above).
6. Now GPP, NPP and R can be calculated. Units of productivity are expressed as energy used e.g. joules $\times$ m^{-2} $\times$ day^{-1}

Catching small motile animals

These are more problematic as they move around, so how do we count small animals? Obviously they have to be caught first. Make sure you can identify the insects you are likely to catch – have a key to help you.

WARNING: Under no circumstances should any animal be stressed or killed during any investigation – there are humane ways to catch and count small animals. Safe, harmless techniques that can be used to catch insects include:

1. Pitfall trap.
2. Sweep nets.
3. Tree beating.

WARNING:

- Make sure there are no venomous organisms in your local area.
- DO NOT handle the insects directly – move the insects with tweezers or a pooter.

Pitfall traps

The pitfall trap is ideal for catching insects and small crawling animals that cannot fly away. Insects can be attracted by decaying meat or sweet sugar solution and will fall into the trap.

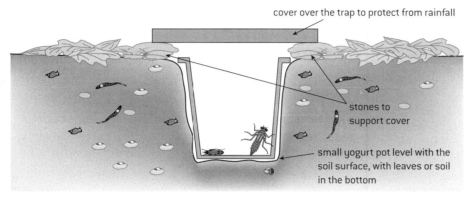

cover over the trap to protect from rainfall

stones to support cover

small yogurt pot level with the soil surface, with leaves or soil in the bottom

▲ **Figure 3.12** Pitfall trap for collecting small insects

Several of these traps can be placed around the study area. They should be checked at regular intervals (every six hours) and the species and number of that species recorded.

WARNING:

- DO NOT put any fluid in the bottom of the trap – you do not want to kill the insects.
- DO NOT leave the traps unchecked for more than 24 hours.

Sweep nets

Sweep nets of various sizes can be swept through grasses at various heights in order to catch insects.

These can then be emptied into a large clear container and the species and numbers recorded.

▲ **Figure 3.13** Sweep net

▲ **Figure 3.14** Tree-beating to collect insects in branches

Tree beating

This method can find insects in tree branches. Simply place a catching tray beneath a tree branch and gently tap the branch. The tray will catch anything that falls from the tree and you can log the species and their numbers.

Night-flying moths will be attracted to a light; place a white sheet behind a light and the moths will settle on this for you to observe.

Small insects and invertebrates can be caught with a pooter – a small jar with two tubes attached. You suck gently on one tube and the animal is pulled into the jar. You cannot swallow it as there is gauze at the end of the mouthpiece tube!

suck here

rubber tubing

rubber bung

gauze tied on to prevent animal entering

animal in here

specimen tube

▲ **Figure 3.15** A pooter

Kick sampling

The organisms of most interest in a freshwater stream will be the invertebrates and the most efficient way to catch them is through kick sampling.

Another simple technique:

- Place the sweep net downstream from you.
- Shuffle your feet into the streambed for 30 seconds.
- Empty the contents of the net into a tray filled with stream water.
- Use a pipette to sort the various insects into small plastic cups and record your results.
- Repeat three times to ensure good results.

Measuring abundance

Mobile animals: Lincoln index (capture, mark, release and recapture)

Sessile or slow moving animals can be counted as individuals, for example, limpets and barnacles. More mobile animals are harder to assess and the **Lincoln index** is used to estimate the population size of animals which move about or do not appear during the day.

Method:

1. Establish the study area.

2. Capture a sample of the population. The actual method of capture will depend on the size of animals; you can take your pick from the methods discussed earlier.

3. Mark each of the organisms captured and record how many you have marked: this must be done in a non-harmful way that does not expose them to higher predation levels than non-marked individuals. For example, dog whelks on a rocky shore or woodlice in a woodland can be marked with a spot of non-toxic subtle coloured paint (nothing bright).

4. Release the captured individuals back into the environment and allow sufficient time to remix with the population.

5. Take a second sample in the same way as the first. Count the number of organisms captured in this sample and count how many of them are marked. At least 10% of the marked sample should be recaptured if the estimate is going to be fairly accurate.

Assumptions made are that:

- Mixing is complete, that is, the marked individuals have spread throughout the population.

- Marks do not disappear.

- Marks are not harmful nor do they increase predation by making the individual more easily seen.

- It is equally easy to catch every individual.

- There are no immigration, emigration, births or deaths in the population between the times of sampling.

- Trapping the organisms does not affect their chances of being trapped a second time.

Lincoln index formulae

$$\frac{m_2}{n_2} = \frac{n_1}{N}$$

OR

$$N = \frac{n_1 \times n_2}{m_2}$$

Where

n_1 = number of animals first marked and released

n_2 = number of animals captured in the second sample

m_2 = number of marked animals in the second sample

N = total population (the figure you are after)

Plant abundance

There are a number of ways of assessing plant species abundance.

- Density: mean number of plants per m².

- Frequency: the percentage of the total quadrat number that the species was present in; may also be measured within the quadrat.

- Percentage cover: because plants spread out and grow percentage cover is often measured instead of individual numbers. This is an estimate of the coverage by each species and it sometimes helps if the quadrat is divided up for this. Species may overlap or lie in different storeys in a forest, so the percentage cover within a quadrat may be well over 100% or much less if there is bare ground.

  The percentage cover can be estimated by comparing the sample area with figure 3.16 and then it can be graded on a scale from 0 to 5 on the ACFOR scale by using figure 3.17.

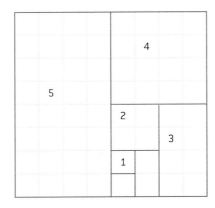

▲ **Figure 3.16**

Percentage cover (%)	ACFOR scale	Score
>50	Abundant	5
25–50	Common	4
12–25	Frequent	3
6–12	Occasional	2
<6, or single individual	Rare	1
absent		0

▲ **Figure 3.17** Percentage cover scales

Possible fieldwork investigations

It is good to start with a walk around your local area or school grounds and observe what is there.

- Is there a playing field?
- Is there a footpath on soil rather than concrete?
- Does the ground slope?
- Is it more shady or more moist in one area than another and what difference does that make to the type and number of species living there?

Measure changes in abiotic factors and/or biotic factors in a local ecosystem

There are a number of changes that can take place.

1. Over space: an environmental gradient – which is a trend in one or more abiotic and/or biotic components of an ecosystem.

2. Over time: short-term diurnal cycles (day and night) or long-term changes (succession).

3. Changes due to human activity: sewage effluent outfall, intensive agriculture.

Investigate the changes that occur along an environmental gradient in your local area

This could be used to practise any of the IA skills and be included in the PSOW but would not be suitable for a full IA unless there is an environmental issue and the societal element is strong.

The environmental gradient you select will depend on where you are doing the investigation. You could investigate:

1. up or down a hill slope

2. along a stream

3. travelling away from a river or some other linear feature (road)

4. in a line away from the sea or lake shore (from shallow water to land)

5. through a woodland area from edge to centre.

> **Tip**
>
> Make sure you select a research question for your IA investigation that can meet the criteria of the IA. IA criteria should guide you in deciding your IA research question:
>
> - Identifying context – only suitable if you can identify an environmental issue and societal link
> - Planning
> - Results, analysis and conclusion
> - Discussion and evaluation
> - Application
> - Communication.
>
> The investigations here would help you gain the skills you need to carry out an IA but may not be enough to gain high marks in your IA unless you can link your research question to an environmental issue (local or global or both), evaluate it in depth and justify and evaluate an application or solution to the environmental issue. Be very sure that you have read and understand the IA criteria.

This setting is ideal as you can measure changes from the top of one hill, into the valley and up the other side.

You could also measure changes down the valley.

▲ **Figure 3.18** Wooded and open valley

To investigate the changes you need to:

1. Set up transect lines (at least three). In this investigation you cannot use random sampling as you are expecting to see a regular change.

2. In an investigation like this you can either do systematic sampling (depending on the length of the slope, e.g. every 25 m). Alternatively you can measure the slope, and divide it into equal sections.

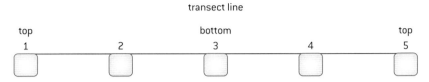

Quadrats placed on the same side of the transect line in the same position every time

▲ **Figure 3.19** Possible sampling of the slope

Abiotic factors which could be measured at each of the sample points along the transect line include soil factors and temperature, light intensity and wind speed.

All measurements are taken at the set sampling points along the transect lines.

Soil infiltration rate

This must be done in the field using the following method (AWAY from where the soil samples are taken).

1. Select a suitable spot to measure drainage – it should be flat.

2. Take a short section of sturdy plastic tubing (drainpipe).

3. Knock the tubing about 15 cm into the soil.

4. Pour a set amount of water into the tube and time how long it takes for the water to drain away completely:

    a. if the soil has a high clay content or is compacted drainage will be poor and it will take longer for the water to drain away (poorly drained)

    b. if the soil is sandy the water will drain away quickly (well drained).

▲ **Figure 3.20** Measuring soil infiltration rate

Soil texture, pH and mineral, organic and water content

The above factors can all be measured most accurately back in the lab but it is essential to take samples carefully. To collect soil samples at each stop:

1. Select suitable sample points – maybe as shown below.

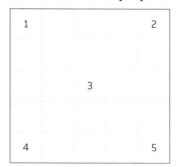

2. Sink a soil auger (there are many types) to a set depth.

3. Remove all soil from the auger.

4. Place the soil in a ziplock bag and remove as much air as you can.

5. Label the sample very clearly to indicate exactly where the sample was taken from. For example Transect 1 Site 1 (bottom of hill), sample 1. The label can be on the actual bag OR in the bag written on plain paper in pencil.

Temperature, light intensity and wind speed

For these measurements to be accurate and comparable along the gradient the measurement must be taken at the same TIME at all sample points. This means that you need a team of people (maybe the whole class) taking these measurements simultaneously at some point in the investigation. Each measurement should be taken five times as described in the previous section.

Biotic factors that could be measured at each of the sample points along the transect line

The biotic factors that could be measured are more limited and it is probably better to stick to plants in this type of investigation. You could measure:

1. Plant species abundance (see page 17)
    a. density
    b. frequency
    c. percentage coverage

2. Species diversity
    a. You can use frequency information to calculate the Simpson diversity index for each quadrat:

    $$D = \frac{N(N-1)}{\sum n(n-1)}$$

    where

    D = Simpson diversity index

    N = total number of organisms of all species found

    n = number of individuals of a particular species

Data collection

This type of investigation requires very clear quantitative data tables – it is a good idea to try and plan your data collection tables ahead of time. Always write in pencil when recording fieldwork data – pencils work in the rain!

Qualitative data will also be very important – describe the weather, take pictures along the transect lines and of the surrounding areas.

Data presentation

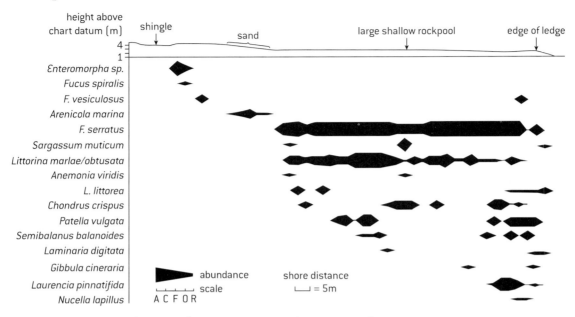

▲ **Figure 3.21** Kite diagram showing one way of presenting data

Kite diagrams are an excellent way to show the spatial distribution of a plant species, especially along an environmental gradient, succession or zonation.

Assess biodiversity using the Simpson diversity index

How diverse is the local ecosystem?

This could be used to practise the following IA skill:

- Planning – you will have to plan the method for this.

▲ **Figure 3.22** Possible ecosystem for biodiversity measurements

1. Select your study area and decide on a sampling strategy and quadrat size. In the photograph shown here random sampling with quadrats that are 0.5×0.5 m would be suitable. (see page 23 for the method to select sample points).

2. Design a data recording table to ensure you record all the relevant information (in a real investigation you will need more than four quadrats).

Quadrat number	Species	Frequency of species	Total (n)
1			
2			
3			
4			

3. Go out and collect the data.

4. Calculate the Simpson diversity index for your study area.

5. It is good if you can find a number of contrasting areas and compare the diversity. For instance, the landscape below gives three contrasting areas.

▲ **Figure 3.23** Possible landscape for comparing biodiversity in three areas

Investigate changes in abiotic and biotic factors over time

Changes over time can occur over vastly different time spans, e.g. day–night cycles, seasonal changes or long-term succession. But such investigations require sampling/measuring on a regular basis over a suitable period. 'Before' and 'after' measurements are not enough. When measuring long-term changes you should do all measurements at the same time of the day to prevent interference by day–night cycles. For example, measuring temperature variation in a wood and a field will only give you meaningful results if you measure both at the same time on the same days.

This could be used to practise the following IA skills:

- Planning
- Results, analysis and conclusion

Investigate the impact of an abiotic factor on a biotic factor

This is a very straightforward, traditional 'wet lab'. You can investigate the impact of a wide range of abiotic factors on terrestrial or aquatic plants or even seeds. Some of these could also count as an IA as the changes in the abiotic factors could be caused by human activities.

> This could be used to practise any of the IA skills but would not be suitable for a full IA unless there is a good environmental issue and a societal link:
>
> - Identifying context – only if you can identify an environmental issue and societal link
> - Planning
> - Results, analysis and conclusion
> - Discussion and evaluation
> - Application
> - Communication

Investigate the impact of abiotic factors on the rate of photosynthesis in terrestrial plants

The choice of plants will depend on the area in which you are conducting the investigation. Fast-growing plants are needed and mint plants are ideal; there may be better ones in your environment so check it out. The easiest ways to measure the rate of photosynthesis in terrestrial plants are to measure the increase in stem length of the plant or the increase in the number of leaves on each stem.

Factor to change (independent variable)	Suggested increments
Water pH (Possible environmental issue here)	pH3, pH5, pH7, pH9 and plain water (depending on the plants' usual tolerance)
Nutrients (fertilizer)	Liquid fertilizer/100ml of water None, 5 drops, 10 drops, 15 drops and 20 drops. This will need to be checked on the bottle.
Salinity (Possible environmental issue here)	Grams of salt/100 ml of water None, 5, 10, 15 and 20 (depending on the plants' usual tolerance)
Water amount	ml/day None, 20, 40, 60 and standing in water (depending on the plants' usual tolerance)
Light wavelength	Different colour filters Clear, red, green, blue and yellow (depending on what you can get hold of)
Light intensity	Various wattages of light bulbs or differing distances from the light source
Light duration	Number of hours of light Total darkness, 6, 12, 18 and permanent light

▲ **Figure 3.24** Abiotic factors that you can use as the independent variable

The increments will vary with different plants; you need to do some investigation to establish the normal range for the plant you are using.

To ensure a valid set of data you will need five trials per independent variable increment and five increments of the independent variable, e.g. for water pH you should have five plants/stems in pH3, five in pH5 and so on.

You should also run the test for 7–10 days to ensure you give the plants time to adjust to the new conditions.

Investigate the impact of abiotic factors on the rate of seed germination

Seeds are relatively easy to acquire and it is a good idea to do a trial run to establish which seeds germinate quickly under ideal conditions (something you can check before you start). It is better to pick large seeds in which it is easy to find the root and shoot. Examples of large seeds that germinate easily are peas and beans.

Factor to change (independent variable)	Suggested increments
Water pH (Possible environmental issue here)	pH3, pH5, pH7, pH9 and plain water
Salinity (Possible environmental issue here)	Grams of salt/100 ml of water None, 5, 10, 15 and 20
Water amount	ml/day This will depend on the size of the container you are using
Light duration	Number of hours of light Total darkness, 6, 12, 18 and permanent light

▲ **Figure 3.25** Possible factors to alter in investigating seed germination

Make sure you have five trials per independent variable and five increments of the independent variable. Record suitable data, e.g. speed of germination, length of root/shoot and qualitative changes.

Investigate the impact of abiotic factors on the rate of photosynthesis in aquatic plants

Again the choice of plants will depend on the area in which you are conducting the investigation and your choices for aquatic plants may be limited. *Elodea* is a good option but may not be available in some areas. To measure the rate of photosynthesis you can count the number of bubbles released per minute.

Factor to change (independent variable)	Suggested increments
Water pH (Possible environmental issue here)	pH3, pH5, pH7, pH9 and tap water (measure pH)
Carbon dioxide concentration	Sodium hydrogen carbonate (g)/100 ml of water: 0, 5, 10, 15 and 20g
Salinity (Possible environmental issue here)	Sodium chloride (g)/100 ml of water 0, 5, 10, 15 and 20g
Water temperature (Possible environmental issue here)	5 °C, 25 °C, 45 °C, 65 °C and room temperature
Light intensity	Various wattages of light bulbs or differing distances from the light source
Nitrate or phosphate loading (Possible environmental issue here)	Liquid fertilizer/ 100ml of water None, 5 drops, 10 drops, 15 drops and 20 drops. This will need to be checked on the bottle

▲ **Figure 3.26** Possible abiotic factors to alter in investigating rate of photosynthesis

To ensure a valid set of data you will need five trials per independent variable increment and five independent variable increments.

Measure the primary productivity and biomass in the local ecosystem

This can be done for both terrestrial and aquatic ecosystems. It is a good idea to make sure you have permission to harvest plant material before you go ahead and start this investigation.

This could be used to practise the following IA skills:
- Planning
- Results, analysis and conclusion

This can be done over space or over time and in both aquatic and terrestrial ecosystems.

You could measure:

- Differences in biomass/primary productivity around your school.
- Changes in biomass/primary productivity over a year. You could collect and measure biomass/primary productivity of your study area once a month for a year.

You will have to select

1. an appropriate sampling strategy
2. an appropriate quadrat size.

The method for measuring biomass is given on page 61.

Questionnaires

Questionnaires allow for the collection of up-to-date data that is specific to the purpose of the study. Well-designed questionnaires allow for graphical displays and statistical analysis, although in some cases you may need to weight the response for really useful data.

Questionnaire design

Some basic rules for designing a questionnaire are:

1. Keep it short, no more than 5–7 questions. People may be in a hurry so a short simple questionnaire allows them to continue with their daily routine without too much interruption.

2. Keep it simple. You might know everything there is to know about what you are investigating but your target audience may not. Avoid technical terms.

3. Stick to closed questions; they are easier to analyse and display graphically. Closed questions are where you give a limited number of options to a particular question.

> How do you get to school?
>
> | | |
> |---|---|
> | Walk | |
> | Cycle | |
> | School bus | |
> | Public transport | |
> | Car | |
> | Other | |
>
> **Figure 3.27** Example of a closed question
>
> Tick the appropriate response and move quickly on to the next question.

4. Question order

   a. Start with a screening question to establish whether or not the person should complete the questionnaire. You may only want people who are resident in the country of study or you may only want tourists.

   b. Move on to simple questions. You may need to know age group or education level. These are often the independent variable aspect of your survey.

   c. Then ask the harder questions – probably the ones that involve the substance of your study (maybe the dependent variable).

Key term

A **questionnaire** is a series of questions with a limited set of responses designed to obtain information about a particular topic.

5. Do not use biased questions that will push the respondent in a particular direction.

6. If you have sensitive questions like age, income or education level make sure you have closed questions with categories for responses. This is often the case with the independent variable questions (4b) but may be true of other questions so be careful.

How old are you?

Under 21	
21–40	
41–60	
Over 60	

Figure 3.28 Example of how to tackle sensitive questions

The groupings you use will depend on the specific situation.

WARNING: Make sure the categories do not overlap at all.

7. Run a pilot with friends and family to make sure it gives you the information you need.

8. Make any necessary adjustments based on the pilot.

9. Make arrangements for the respondent to complete the questionnaire anonymously. In face-to-face interviews give the respondent the questionnaire and let them place it in an envelope.

10. ALWAYS ask yourself 'does this question relate clearly and directly to my investigation?' If the answer to that question is no remove the question and THINK AGAIN.

Advantages and disadvantages of questionnaires

As with all methods of data collection questionnaires have advantages and disadvantages.

Advantages

1. The questions and responses are standardized so they are reasonably objective.

2. Data collection is quick.

3. You can collect a lot of information in a short period of time, especially if you are doing face-to-face data collection.

Disadvantages

1. Data collection may be quick but good questionnaires take time to design.

2. You are sometimes relying on the memory of the respondents – not always very accurate.

3. If you are not conducting the questionnaire face-to-face you cannot explain the questions if they are not clear to the respondent.

4. Open-ended questions are difficult and time-consuming to analyse effectively. So do not use too many.

5. If the questionnaire is too long there is the risk that responses may be superficial. So keep it short.

6. People may not answer honestly if they feel embarrassed or if they think the response will damage them in some way.

Distribution of the questionnaire

Having designed and piloted the questionnaire you now have to collect the information. There are a number of options for this.

1. Face-to-face interviews

    This involves you going out and actually conducting the survey face-to-face with the target population. This method can be time-consuming but it does tend to produce plenty of data. Where you go to conduct the survey will depend on the topic at hand but generally speaking you can:

    - go to local centres of population (towns, villages, cities)
    - go to tourist sites, beaches and parks
    - visit local shopping centres
    - use your school population – students, teachers and parents.

    You may need to seek permission for some of these options, so check first.

2. Email/online

    If you are using your school population you may be able to get access to the school's list of email addresses for students and staff and send your survey out by email. Many websites now have a questionnaire facility so you could send the survey out as an online survey.

3. Mail – very traditional and rather slow. Not recommended.

Sampling methods

Generally speaking it is impossible to complete your questionnaire for the whole population; therefore it is necessary to ask a representative sample of the population. However, the sample must be unbiased, representative of the whole population and include at least 30 people. More is better and more reliable, but also more time-consuming.

There are three techniques that can be used to achieve an unbiased sample for questionnaires.

1. Systematic
2. Stratified
3. Random

Systematic sampling

This is sometimes called 'nth' sampling – you simply ask people at a given interval, for example, every 5th person. The interval will depend on where you are and whether or not the area is busy. This method is very

Key term

Sampling is a statistical technique that allows you to obtain representative data from a small portion of the whole population.

straightforward and the most useful method if you are dealing with an unknown population – for example, asking people on the street or in a mall.

1. Select an interval – ask every 5th person that passes you.

2. Select a strategic sampling position.

3. Stop every 5th person that passes you and ask them if they mind completing your questionnaire.

4. Conduct your questionnaire with that person.

5. Repeat this process until you have a minimum of 30 completed questionnaires.

Stratified sampling

This is where you are targeting particular groups for the questionnaire; you may want to ask just men, just women or a particular age group. This opens up two possibilities.

- Do you want an even number of respondents in each of your groups

- Do you want a number of respondents in each category that is representative of the population as a whole. This may be hard to achieve.

The choice is yours.

Using the earlier example:

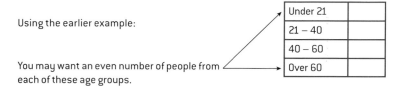

Using the earlier example:

You may want an even number of people from each of these age groups.

Under 21	
21 – 40	
40 – 60	
Over 60	

In this case you can stop every person that passes you and ask which age group they fit into. Only continue with the questionnaire if they are in the age group(s) that interests you. This should be repeated until you have a minimum of 30 completed questionnaires in total.

Random sampling

This is where every member of the population has an equal chance of being selected and involves the use of random numbers tables/generators (GDC, random.org). With questionnaires this can only be done if you have access to a list of the whole population. For example, you may be conducting the questionnaire for the whole school, a particular year group or all parents.

1. Take a list of the entire population and allocate a number to each member of the population.

2. Generate a random number.

3. Find that number on your list – this person will be one of your respondents.

4. Repeat this until you have a minimum of 30 names to conduct the questionnaire with.

Possible uses of a questionnaire

Questionnaires are a tool which could form part of your investigations. The following suggested investigations show how you could use questionnaires to generate data; they are not full IAs.

Tip

Use random.org to generate a random number:

a. Go to www.random.org

b. Go to the 'True random number generator' and set the minimum number to 1 and the maximum number to the last number on your list.

c. Click 'generate' – this will give you a number.

1. Investigate the relationship between gender (or age) and environmental attitudes

Research question: Is there a relationship between gender (or age) and environmental attitudes?

This could be used to practise the following IA skills:

- Planning – if you use this as an example then plan your own investigation
- Results, analysis and conclusion
- Discussion and evaluation
- Application
- Communication

How to tackle this research question

1. Select a number of environmental issues that are covered in the ESS syllabus.

2. Design a questionnaire so that each question has three responses:

   a. Ecocentric based

   b. Anthropocentric based

   c. Technocentric based

There are three main environmental value systems (EVSs):

- The **ecocentric** worldview puts ecology and nature as central to humanity and emphasizes a less materialistic approach to life with greater self-sufficiency of societies. It is life-centred – respects the rights of nature and the dependence of humans on nature so has a holistic view of life which is Earth-centred. Extreme ecocentrists are deep ecologists.

- The **anthropocentric** worldview believes humans must sustainably manage the global system. This might be through the use of taxes, environmental regulation and legislation. It is human-centred –humans are not dependent on nature but nature is there to benefit humankind.

- The **technocentric** worldview believes that technological developments can provide solutions to environmental problems. Environmental managers are technocentrists. Extreme technocentrists are cornucopians.

3. Ask your target audience what type of solution they think is the best – only allow them to select one solution.

Collecting the data: the questionnaire

The responses on this questionnaire are given a letter (e, a, or t) to make it easier for you to understand. Do not do this for the questionnaire you use.

1. Gender (this could be the independent variable)   Male ☐   Female ☐

2. The photograph shows one of the impacts of global climate change. This is considered by some to be a major environmental issue – which one of the following solutions do you think is most suitable?

We must educate people to encourage the reduction of greenhouse gases (GHGs) – use public transport, reduce electricity consumption, change diets etc. (e)	
We must regulate the production of GHGs through legislation and taxes. (a)	
We must look to technology for solutions – renewable energy, scrubbers, hybrid cars etc. (t)	

3. Human population growth is significant. Most people agree that this will cause problems – what is the solution?

Further scientific research is needed to ensure we can increase space, food production, water supply and resources. (t)	
It does not matter if people become less materialistic and more self-sufficient. (e)	
Policies such as China's 'One child policy' should be employed to bring population growth under control. (a)	

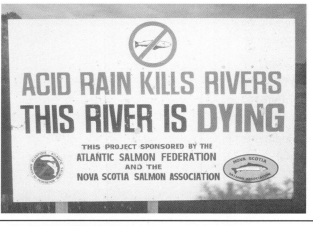

4. Acid deposition is a serious problem in some areas of the world. How should we deal with this issue?

We must educate people to encourage them to reduce the combustion of fossil fuels that cause acid deposition. (e)	
We must use legislation and impose taxes in order to reduce the production of the gases that cause acid deposition. (a)	
We must look to technology for solutions – renewable energy, scrubbers, hybrid cars etc. (t)	

5. The Great Pacific Garbage Patch (GPGP) is a mass of plastic in the middle of the Pacific Ocean. How can we avoid adding more plastic to it?

Clean up the GPGP. (a)	
Reduce, reuse and recycle. (t)	
Raise awareness (through education) of the concept of biorights and the need for humans to self-regulate consumption of plastics. (e)	

- These are possible questions on a few of the topics you study – the aim is to give you some ideas. Make your own questionnaire to suit your area and your target audience.

- You do not want to ask too many more questions; 5–7 is an ideal number for a questionnaire.

- Make sure you mix up the responses so there is no clear pattern for respondents to follow.

- You should adjust the responses to suit your audience, especially if you are going to ask the general public.

- Use your own pictures to suit your area and your audience. Pictures are not essential but it gives people an idea of what you are talking about.

- Other possible issues that you could use:
  - loss of biodiversity
  - eutrophication
  - soil degradation
  - ozone depletion
  - tropospheric ozone/photochemical smog
  - resource depletion
  - solid domestic waste.

Once you have designed the questionnaire you select a sampling technique and go out and gather your data.

Presenting the data

1. The first step is to collate the questionnaire data. This is quite simple – separate the questionnaires into piles according to the independent variable, in this case gender or age.

2. Using a blank questionnaire use a five-bar tally system to record how many people responded to each question in each of the response categories.

Example of collation for females for question 2.

The photograph shows one of the impacts of global climate change. This is considered by some to be a major environmental issue – which one of the following solutions do you think is most suitable?	
We must educate people to encourage the reduction of GHG's – use public transport, reduce electricity consumption, changing diets etc	### ### ###	
We must regulate the production of GHG of through legislation and taxes.	////
Look to technology for solutions – renewable energy, scrubbers, hybrid cars etc	/

This is the 5 bar tally system and it makes questionnaire collation very easy

Figure 3.29

3. This will give you a table that summarizes the collated data (similar to the one below).

	Ecocentric		Anthropocentric		Technocentric	
	Male	Female	Male	Female	Male	Female
Global climate change	5	7	3	5	7	5
Human population growth	4	8	4	4	8	4
Acid deposition	3	9	3	5	9	3
GPGP	2	5	2	6	5	2

Figure 3.30

(Note – these are purely fictitious numbers.)

4. These numbers can then be used to produce a graph – this is one of many possibilities:

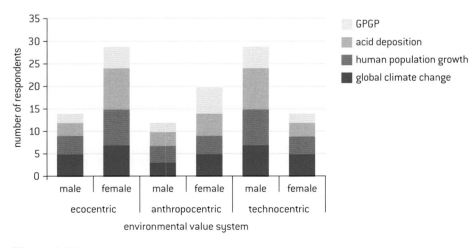

Figure 3.31

2. Environmental attitude survey

This could be used to practise the following skill:

- Planning – if you use this as an example then plan your own investigation

There are many quotes throughout sub-topic 1.1 in the Oxford ESS course companion – they are all largely ecocentric in their sentiment. You could use a number of them to assess people's attitude to the environment (see below).

Statement	Agree	Disagree
'Let us a little permit Nature to take her own way; she better understands her own affairs than we.' Michael Montaigne		
'For the first time in the history of the world, every human being is now subjected to contact with dangerous chemicals, from the moment of conception until death.' Rachel Carson		
'There are no passengers on Spaceship Earth. We are all crew.' Marshall McLuhan		
'We shall require a substantially new manner of thinking if mankind is to survive.' Albert Einstein		
'Every creature is better alive than dead, men and moose and pine trees, and he who understands it aright will rather preserve its life than destroy it.' HD Thoreau		
'We do not inherit the earth from our ancestors, we borrow it from our children.' Attributed to Chief Seattle		
'The system of nature, of which man is a part, tends to be self-balancing, self-adjusting, self-cleansing. Not so with technology.' EF Schumacher		
'Your grandchildren will likely find it incredible– or even sinful – that you burned up a gallon of gasoline to fetch a pack of cigarettes!' Paul MacCready Jnr		
'Don't blow it – good planets are hard to find.' Quote in *Time* magazine		
'A thing is right when it tends to preserve the integrity, stability and beauty of the biotic community. It is wrong when it tends otherwise.' Aldo Leopold		

Figure 3.32

You do not have to use these quotes; you can use any others that you prefer. Go to the internet and get your own quotes – maybe you can find some that show anthropocentric or technocentric attitudes. Great websites for quotes include:

http://www.brainyquote.com/

https://www.goodreads.com/quotes.

Follow these links and put your area of interest in the search box at the top of the page.

3. Environmental value systems and natural resources – a bipolar analysis

Another type of questionnaire that can be used to collect data is a bipolar questionnaire. A series of questions can be asked and the respondent rates their feelings or ideas on a sliding scale. Again you could do this type of questionnaire to compare the differences between ages, genders, income groups etc. Or you could just gather the overall data to check opinions in a population as a whole.

This could be used to practise some of these skills:

- Planning – if you use this as an example then plan your own investigation
- Results, analysis and conclusion

Collecting the data: the bipolar questionnaire

Which one of the three approaches do you think will work with the following natural resources?	ECOCENTRIC People can be educated to see the environment holistically. People must exercise self-restraint and be • less materialistic • self-sufficient.	ANTHROPOCENTRIC People can manage the environment sustainably through • taxes, environmental regulation and legislation • debate to reach a consensual, pragmatic approach to solving environmental problems.	TECHNOCENTRIC Technological developments can provide solutions to environmental problems and improve the lot of humanity. This can be done through scientific research to form policies and understand how systems can be controlled, manipulated or exchanged to solve resource depletion.
Fossil fuel depletion			
Water shortages			
Soil depletion and desertification			
Precious metals/gemstone depletion			
Biodiversity loss			
Landscape degradation			
Deforestation			

Figure 3.33

You could support these ideas with pictures to help your respondents visualize the problems associated with the loss/excess use of these natural resources.

Once you have designed the questionnaire you select a sampling technique and go out and gather your data.

To save paper you could have a single questionnaire and fill in the responses on a digital version. This will also mean you do not have to collate the data as you will already have everything in your digital version. WARNING – this method requires a lot of concentration, as it is easy to get distracted and forget.

Presenting the data

1. Again the first step for the paper version is to collate the questionnaire data. If you are gathering general data for the population you do not need to separate into piles (if you have an independent variable you are interested in then you do have to).

2. Use a blank questionnaire and the five-bar tally system to record how many people responded to each question in each of the response categories.

3. This data can be presented as a bar graph, as in the previous example or it can be shown in other ways.

4. Below is a bar graph (again from completely fictitious data) to show total number of responses in each category.

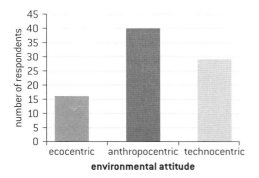

Figure 3.34

5. With graphs you need to spend time to find the most appropriate way to present your data, a way that assists analysis.

For all investigations remember you have to work within the IB rules of animal experimentation.

These activities could be part of your PSOW and give you the skills you need for your IA.

Topic 1: Foundations of environmental systems and societies

1. Investigating systems: Set up an aquatic or terrestrial ecosystem in a bottle

This could be used to practise the following IA skills:

- Planning

(REMEMBER: NO ANIMALS SHOULD BE HARMED)

A terrarium is a mini indoor garden that models a closed system. Check out Mr. Latimer's ecosystem in a bottle which has been watered once in 53 years and has been a closed system for 40 years: http://www.dailymail.co.uk/sciencetech/article-2267504/The-sealed-bottle-garden-thriving-40-years-fresh-air-water.html

To start this investigation, create a systems diagram to show the inputs and outputs of the terrarium. The inputs will allow you to make a list of equipment you will need.

Materials

1. Plants:
   a. Choose small plants that you like, that will not outgrow the terrarium and that grow well together – you may need to do some research. The plants selected will depend largely on where you live – mosses and ferns are good.
   b. Terrariums tend to have low light levels as the plants are close together so make sure the plants can tolerate such conditions.
   c. Closed terrariums have high humidity because once the garden is set up it is sealed off, so make sure the plants are tolerant of high humidity.
   d. A terrarium community is hard to balance so pick plants that are easy to grow and inexpensive.
2. Container: there is a wide range of containers that you can use. The most usual is a glass jar with a tight-fitting lid. Make sure it is big enough for the plants and that there is plenty of room for root development.

Or you could help reduce waste and use old plastic soda bottles as in figure 3.35.

▲ **Figure 3.35** Recycling an old soda bottle for a terrarium

3. Location: decide where you will put the terrarium – once they are up and running there should be no maintenance (if they work).

   a. Make sure there is plenty of indirect light – direct light is too harsh and may cause extreme variations in temperature.

   b. Keep the terrarium in a warm indoor environment where there are no extremes of temperature.

4. Soil: use light potting compost that drains easily. Do some research to find out which soil is best for your plants and a terrarium.

5. Pebbles or gravel: these go at the bottom of the terrarium to provide drainage or for decoration on the top of the soil.

6. Activated charcoal: this keeps the soil fresh.

7. Sheet moss tends to soak up excess water so it is useful to have in the bottom of the terrarium.

Method

1. Clean the container.

2. Line the bottom with layers of:

   a. sheet moss

   b. pebbles or gravel.

3. Add the soil and activated charcoal.

4. Add the plants.

5. Leave it open for a while to settle and establish the right amount of water.

6. Seal the container.

Now you have made a terrarium (a terrestrial ecosystem) do your own research and set up an aquatic ecosystem in a bottle (an aquarium).

2. Investigation of feedback

This is a computer simulation game called Sunny Meadows.

Go to:
http://www.goldridge08.com/flash/fc44/foodchain.swf

Read the ten slides and construct a negative feedback diagram from the information on the slides.

The aim of the game is quite simple – get the highest score you can. This is achieved through stable population numbers.

- Click play game to start.

- Choose view: picture, graph or biomass (HINT – the graph is probably the most helpful if you aim to improve your score).

- Choose game speed: 1× allows you to read the comments as the game progresses – this will help you improve your score.

- Pick starting numbers for:
    - foxes
    - rabbits
    - grass.

  Which one do you think will dictate the stability of the system?

- Let the game run for 50 years, recording your scores on the table below.

- Repeat until you achieve a high score of over 90.

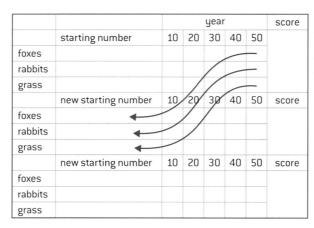

The final numbers of one trial are the new input numbers for the next trial. They will not be exactly the same but get them as close as you can.

		year					score
	starting number	10	20	30	40	50	
foxes							
rabbits							
grass							
	new starting number	10	20	30	40	50	score
foxes							
rabbits							
grass							
	new starting number	10	20	30	40	50	score
foxes							
rabbits							
grass							

▲ **Figure 3.36**

Topic 2: Ecosystems and ecology

1. Build up a food chain for a local ecosystem

This could be used to practise the following IA skills:

- Planning
- Results

You should select an area very close to where you live or go to school because you will need to set pitfall traps.

You will need to:

1. Sample the vegetation in the study area – this does not need to be detailed, you merely need to establish a list of all the primary producers that are in the area. You will need to establish:

   a. an appropriate sampling strategy

   b. an appropriate quadrat size.

2. Set pitfall traps throughout the study area. This will show what insects are in the area and possibly some small amphibians.

3. Whilst sampling the vegetation look for evidence of secondary consumers – faeces, burrows, footprints, scratch marks etc.

4. You can use keys to identify the organisms and you will have to do some research about the organisms' feeding habits.

2. Design your own dichotomous key for six or more organisms in your local ecosystem

Go out and about around the local environment and collect or take a photograph of six or more organisms.

▲ **Figure 3.37** Six organisms

An easy way to create keys is using a mind map such as https://bubbl.us/.

The mindmap in figure 3.38 shows how to make a dichotomous key using the organisms in figure 3.37.

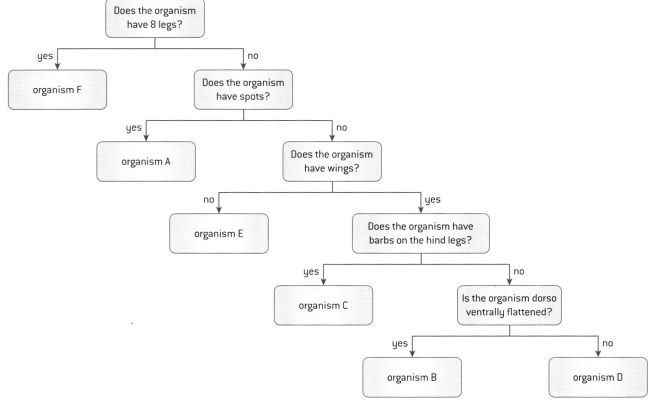

▲ **Figure 3.38** Dichotomous key created using a mind map tool

3. Investigate the efficiency of the Lincoln index

This is a class activity.

1. Take 500 large beans, preferably light in colour, for example, large white lima beans or chickpeas.

2. Place them in a suitable container with a lid.

3. Each student takes a small handful of beans, counts them and marks them in some way (n_1). This number is recorded on the group data table (on the board).

4. The beans are then replaced in the container. The container is shaken vigorously to try and remove some of the marks.

5. Each student then takes a second handful of beans and records the total number of beans in this sample (n_2) and the number of beans that are marked (m_2).

6. Apply the Lincoln index formulae to each of the rows of data.

7. Complete the data recording table on the board.

Student	n_1	n_2	m_2	N
1				
2				
3				
4				
5				
6				
7				

$$N = \frac{n_2 \times n_1}{m_2}$$

n_1 = number of animals first marked and released

n_2 = number of animals captured in the second sample

m_2 = number of marked animals in the second sample

N = total population (the figure you are after)

▲ **Figure 3.39**

4. How diverse are local car parks?

This could be used to practise the following IA skills:

- Planning
- Results, analysis and conclusion

In some locations the local ecosystem may not be suitable for investigations. So you can calculate the diversity of cars in the school car parks or local shopping areas.

1. You will not need a sampling strategy – just go to the car parks and count.

2. You will have to decide how you are going to 'species' the car – colour or make; both work.

3. You can use the table below to record data.

Car park location / name	'Species' of car	Frequency of species	Total (n)
1			
2			
3			
4			

▲ **Figure 3.40**

4. Calculate the Simpson diversity index for the car parks.

Topic 3: Biodiversity and conservation

There is a limited range of investigations that can be done in this topic within the time and resources likely to be available to you, but here are a few suggestions.

Biodiversity is the broad term used to describe the variety of life (on Earth, in an ecosystem or habitat). It has three components (species, habitat and genetic). The easiest one to investigate is species diversity. Your ability to be able to conduct this investigation will be determined by where you live and what environments surround you.

1. Investigate species diversity in deciduous woodland and a grassland area (and two areas close to you)

This could be used to practise the following IA skills:

- Planning
- Results, analysis and conclusion
- Discussion and evaluation
- Communication

There are two options in this investigation.

- You could calculate the Simpson diversity index for each of the areas, in which case you will need to collect information on
  - the number of different species (plants and/or animals)
  - how many individuals of each species.
- You could simply construct a list of all the different species (plants and/or animals) in the two areas and see which has the most variety – you will need to continue sampling until you discover no new species in your quadrats. Maybe do a few extra to make sure you have recorded all species.

In the two areas you will need to:

1. Decide if you are collecting data for both plants and animals.
2. Decide on an appropriate sampling strategy – should be the same for both areas.

   REMEMBER – in a woodland area you will have to vary the size of your quadrat to ensure you include the trees.

3. If you wish to collect data on animals you can use
   a. pitfall traps
   b. sweep nets
   c. tree beating.

REMEMBER

When you are collecting your raw data, make notes about weather conditions or anything else that may affect your results. Take pictures as a record. Remember qualitative data. Data can be presented in a variety of ways; it is up to you.

2. Investigate the differences in species diversity in pools and riffles

This could be used to practise the following IA skills:

- Planning
- Results, analysis and conclusion
- Discussion and evaluation
- Communication

In small meandering streams there are areas of fast moving shallow water (riffles) and areas of slow moving deeper water (pools). These areas have a different collection of invertebrate species in them and you could collect data to compare the species diversity.

The same two options exist for this investigation as did for the last one:

- Calculate the Simpson diversity index.
- Count the number of species present in each of the areas.

To collect the data for this investigation you will need to use kick sampling (see page 64).

3. Evolution/natural selection simulations

There are a number of evolution/natural selection simulations on the internet: http://www.nhm.ac.uk/nature-online/evolution/what-is-evolution/natural-selection-game/the-evolution-experience.html

They are generally a fun way to learn about evolution and natural selection.

This could be used to practise the following IA skills:

- Planning
- Results, analysis and conclusion
- Discussion and evaluation
- Communication

Here are some examples to try. You could also devise your own game to demonstrate natural selection.

1. Habitats
   a. Simulate habitats with different background colours.
   b. Simulate organisms with different coloured beads or pieces of paper.
   c. Give yourself 10 seconds to pick out only the animals that you can see easily.
   d. Record the results.
   e. Double the number of organisms that are left in each colour.
   f. Repeat steps c and d until patterns emerge.
   g. You will see certain colours increasing and others may disappear.
2. Predator–prey relationships
   a. You are the predator.

b. Allocate different coloured beads different characteristics:

     i. sting

     ii. run fast

     iii. poison

     iv. slow and tasty!!

   c. You have 10 seconds to get as many slow and tasty ones as you can – if you grab the wrong one by mistake you have to stop.

   d. Record the results.

   e. Double the number of organisms that are left in each colour.

   f. Repeat steps c and d until patterns emerge.

   g. You will see certain colours increasing and others may disappear.

3. Beak shape and survival

   a. Simulate different beak shapes using household items – spoon, chopsticks, fork, bulldog clip and pin.

   b. Simulate different food sources using items of varying shape and size – raw pasta, peas, M&Ms, string, boiled eggs.

   c. Fill the table with one of the food types and using one of the 'beaks' allow 10 seconds to see how much food can be collected. Count or weigh the food and record the results.

   d. Repeat for each beak type.

   e. Repeat for different food sources with different beak types.

   f. This will show adaptation.

	Amount of food collected (whole number)					Total	Mean
Trials	1	2	3	4	5		
Spoon							
Chopsticks							
Fork							
Bulldog clip							
Pin							

▲ **Figure 3.41** Data table to show amount of raw pasta collected by each beak type

4. Investigate global biodiversity

There are a number of investigations that you can do based on secondary data. Nationmaster http://www.nationmaster.com/ is a good source of information on conservation and biodiversity and there are a number of relationships you could investigate.

Possible independent variables	Possible biodiversity measures (dependent variable)
Level of development	Known mammal species
Mean latitude of a country	Endangered species protection
Population size of a country	Protected areas
Deforestation rates in the country	Threatened species

Some of these are more logically paired than others but that is your choice. Here is one pairing.

Investigate the relationship between number of threatened species and the deforestation rates of a country

This could be used to practise the following IA skills:

- Identifying context
- Results, analysis and conclusion
- Discussion and evaluation
- Application
- Communication

1. Find a list of all countries in the world (they must either be numbered or you must number them):

    a. http://www.listofcountriesoftheworld.com/ (these are numbered)

    b. http://www.countries-ofthe-world.com/all-countries.html (these are not numbered, but you can copy and paste them into an Excel workbook and it will number them for you.)

2. You will have to select a sample of countries (minimum of 30). The most appropriate method here is to use random sampling so use a method to generate 30 random numbers.

    a. Go to http://www.random.org/

    b. Use your GDC.

    c. Use random numbers tables.

    d. Many smartphones generate random numbers.

3. Generate a minimum of 30 random numbers and enter them into a raw data table.

Country	Number of threatened species	Change in forest cover 2000–2005 (1000 ha/year)
India	193	+29
Angola	41	−125

▲ **Figure 3.42**

4. To find the number of threatened species:

    a. Go to http://www.nationmaster.com/

    b. Click on 'categories' and select 'environment'.

    c. Select threatened species.

    d. Scroll down to the world map.

    e. Hover over one of the countries in your table and it will show the number of threatened species. Repeat for all countries generated by your random sampling.

5. To find the change in forest cover 2000–2005 (1000ha/year):

    a. Go to http://rainforests.mongabay.com/deforestation.html

    b. Find the countries on your list and record the change in forest cover 2000–2005 (1000 ha/year).

6. You can process the data by carrying out a statistical correlation test.

7. Draw a scattergraph to show the relationship between the two sets of data.

NOTE: There is a large range of indicators that can be used to assess the level of development (independent variable) of a country but the most reliable for an investigation like this are:

- GDP measured in US$
- Human Development Index (https://data.undp.org/dataset/Human-Development-Index-HDI-value/8ruz-shxu).

If you are using level of development as your independent variable you can use one of these:

- Systematic sampling – every 'nth' country from a ranked list
- Random stratified sampling – taking a set number of countries from each of the development categories:
  - very high human development
  - high human development
  - medium human development
  - low human development

Topic 4: Water and aquatic food production systems and societies

1. Investigation of freshwater withdrawal rates by 30 countries

This could be used to practise the following IA skills:

- Identifying context
- Results, analysis and conclusion
- Discussion and evaluation

1. Find a list of all countries in the world (they must either be numbered or you must number them):

    a. http://www.listofcountriesoftheworld.com/ (these are numbered)

    b. http://www.countries-ofthe-world.com/all-countries.html (these are not numbered, but you can cut and paste them into an Excel workbook and it will number them for you.)

2. You will have to select a sample of countries (minimum of 30). The most appropriate method here is to use random sampling so use a method to generate 30 random numbers:

    a. Go to http://www.random.org/

    b. Use your GDC.

    c. Use random numbers tables.

    d. Many smartphones generate random numbers.

3. Generate a minimum of 30 random numbers and enter them into a raw data table.

4. The data on the amount of renewable freshwater is available at:

    a. http://www.nationmaster.com

    b. categories – environment

    c. select 'total renewable water resources'

    d. find the countries on your list and record the total water withdrawal per capita in cu km / capita / year

    You can find your selected countries in one of two ways.

    - Scroll through the list to find the country.

    - Go to the world map at the bottom of the page and hover over the countries you are looking for.

5. The data for total water withdrawal is at http://www.gapminder.org/data/ – put 'Water' in the search space and you will find a range of water related data that you could investigate.

Country	Amount of renewable freshwater (km³)	OR	Total water withdrawal per capita (m³/inhab/yr)
China	2829.6		486.45
Niger	33.7		182.57

▲ **Figure 3.43**

You could collect data on both these water statistics and compare them.

2. Investigate fishing rates in selected countries

This could be used to practise the following IA skills

- Identifying context
- Results, analysis and conclusion
- Discussion and evaluation
- Application
- Communication

You start with steps 1–3 in investigation 1 above.

4. The data on the marine fish catch is available at:

   a. http://www.nationmaster.com

   b. categories – environment

   c. select 'marine fish catch'

   d. find the countries on your list and record the fish catch in million tons

   You can find your selected countries in one of two ways.

   - Scroll through the list to find the country.
   - Go to the world map at the bottom of the page and hover over the countries you are looking for.

5. Alternatively you could use http://en.wikipedia.org/wiki/Fishing_industry_by_country which gives statistics for total marine catch (tons) – fish, shellfish etc.

Country	Marine fish catch (million tons)	OR	Total marine catch (tons)
Peru	8.26		9,416,130
India	2.24		6,318,887

▲ **Figure 3.44**

3. Investigate people's attitudes and EVSs to problems caused by exploitation of the oceans' resources

This could be used to practise the following IA skills:

- Identifying context
- Results, analysis and conclusion

One possible way to investigate this and assess people's approach to environmental management is to use a bipolar analysis (see section 3c. Questionnaires). If you remove the middle ground in this investigation (anthropocentric) people are forced to decide between the two extremes.

There are a number of issues associated with our exploitation of the oceans and these are just a few.

- Harvests are too large of many species.
- Many animals are suffering due to falling fish levels – penguins, seals.
- Some species are harvested to make pet food.
- Shark's fin soup.
- Mining the sea bed.
- By-catch – sharks, turtles and other marine organisms are caught in the nets and die.

WARNING

You must be careful not to be judgmental in this investigation. You should allow the respondents to complete the bipolar questionnaire on their own so they can make their own choices without worrying what is the 'right' answer.

Have this table on a separate sheet for respondents to view with the photos.

I am technocentric and I believe whatever problems we cause, we can solve them.	I am ecocentric and I believe we need the Earth more than it needs us.
<ul><li>We are the Earth's most important species, we are in charge.</li><li>There will always be more resources to exploit.</li><li>We will control and manage these resources and be successful.</li><li>We can solve any pollution problem that we cause.</li><li>Economic growth is a good thing and we should always keep the economy growing.</li></ul>	<ul><li>The Earth is here for all species.</li><li>Resources are limited.</li><li>We should manage growth so that only beneficial forms occur.</li><li>We must work with the Earth, not against it.</li></ul>

OVERFISHING 	I am technocentric on this issue	SHARK'S FIN SOUP 	I am technocentric on this issue
	I am ecocentric on this issue		I am ecocentric on this issue

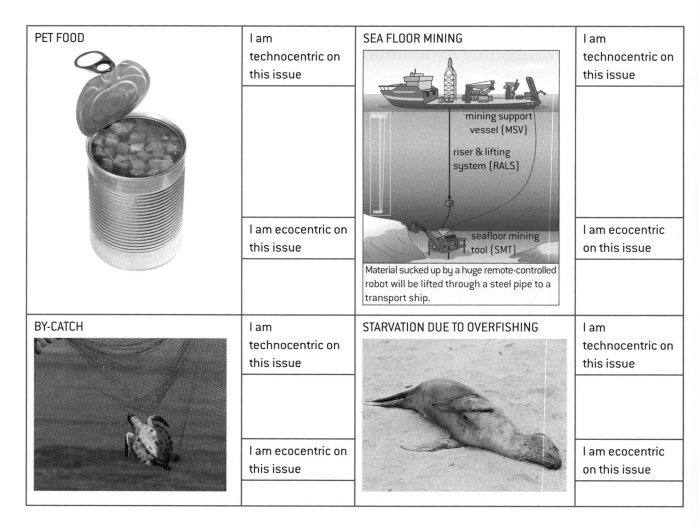

PET FOOD	I am technocentric on this issue	SEA FLOOR MINING	I am technocentric on this issue
	I am ecocentric on this issue	Material sucked up by a huge remote-controlled robot will be lifted through a steel pipe to a transport ship.	I am ecocentric on this issue
BY-CATCH	I am technocentric on this issue	STARVATION DUE TO OVERFISHING	I am technocentric on this issue
	I am ecocentric on this issue		I am ecocentric on this issue

4. Measure the biochemical oxygen demand in a variety of water sources

Biochemical (or biological) oxygen demand (BOD) is a measure of the amount of dissolved oxygen required to break down the organic material in a given volume of water through aerobic biological activity (by microorganisms). It is an indirect measure of organic material in water – dead plants and animals, manure, or even food.

> This could be used to practise the following IA skill:
>
> • Identifying context

Select a range of water sources – various taps, streams, water fountains, lakes etc. The variety of sources will be dependent on where you are and what water sources are available.

At each water source:

1. Prepare eight water collection bottles and make sure they are clean – NO DETERGENT in them.

2. Label the water bottle.

    a. Bottle 1: Tap in school; Bottle 2: Tap in school

    b. Bottle 1: Water tank; Bottle 2: Water tank

    c. Bottle 1: Local lake; Bottle 2: Local lake

    d. Bottle 1: Local river; Bottle 2: Local river.

3. Collect two identical samples of water from each of the water sources.
   a. Seal Bottle 1 immediately (with no air space in the bottle).
   b. Use a probe to measure the dissolved oxygen content of the second bottle. Record the results (mg dissolved oxygen l^{-1}).
4. Place Bottle 1 in an incubator (20 °C) for five days.
5. After five days use the probe to measure the dissolved oxygen content of Bottle 1.
6. BOD = Day 5 reading – Day 1 reading (mg dissolved oxygen l^{-1}).

Examples of BOD values

Source of pollutant	BOD (mg dissolved oxygen l^{-1})
Unpolluted river	<5
Treated sewage	20–60
Raw domestic sewage	350
Cattle slurry	10,000
Paper pulp mill	25,000

▲ **Figure 3.45**

5. Investigate the impact of nitrates or detergents on the growth of pondweed (duckweed)

This could be used to practise the following IA skills:

- Identifying context
- Results, analysis and conclusion
- Discussion and evaluation
- Application
- Communication

The ability to do this investigation is dependent on where you live and the availability of duckweed. Duckweed is a small, light green, freshwater plant. It has a number of advantages for this investigation, as it:

1. grows well in sunlight or shade
2. is tolerant of a wide range of pH (4.5–7.5)
3. is tolerant of the cold
4. grows quickly – especially in the presence of nitrogen and phosphorus.

Ideally you should be able to let this investigation run for a month or so (depending on temperatures and time of the year).

Method

1. Take five clean large bowls (at least 15 cm diameter).
2. Place the same amount of water in each bowl (from the same source). The amount will depend on the size of the bowls.

3. Label the bowls clearly:
   a. Water
   b. Water + 10 grams of washing powder
   c. Water + 20 grams of washing powder
   d. Water + 30 grams of washing powder
   e. Water + 40 grams of washing powder

   The amount of washing powder will depend on the amount of water you use and the size of the bowls.

4. Introduce five lobes of duckweed to each bowl.
   a. Take a picture.
   b. Place a gridded quadrat over the bowl and assess what percentage of the bowl is covered with duckweed.

5. Leave in a warm sunny place and check every week.
   a. Take a picture.
   b. Place a gridded quadrat over the bowl and assess what percentage of the bowl is covered with duckweed.

NOTE: As the aim of this investigation is to measure the impact of phosphorus on the growth of duckweed, make sure the washing powder contains phosphorus (some do not). You will notice that this does not follow the 5 by 5 rule. There are 5 increments but there are not 5 trials per increment.

You could also do this with liquid nitrogen fertilizer.

Week		Water + different amount of washing powder				
		No washing powder	10 grams of washing powder	20 grams of washing powder	30 grams of washing powder	40 grams of washing powder
1	% of the bowl covered with duckweed					
2						
3						
4						
5						
6						
7						
8						

▲ **Figure 3.46** Sample data collection table

6. Investigate the impact of different surface material on the rate of infiltration to assess the possible impacts on flooding

This could be used to practise the following IA skills:
- Identifying context
- Results, analysis and conclusion
- Discussion and evaluation

Infiltration is the process in which surface water enters the soil. Many things affect the rate of infiltration:.

- Soil texture: sandy soil has much higher infiltration rates than clay soils.
- Preceding conditions: if the soil is already soaked due to previous water input then the infiltration rate will be slow.
- Porosity of the surface material: obviously concrete is not porous and so will not allow any infiltration; vegetated areas will have quite different infiltration rates.

To do this investigation you need to have access to a variety of different surfaces – look around you as you move around school, your home area or travel between the two.

Method

1. Select a minimum of five different surfaces – more if possible.

2. In each area select five places where you can carry out your infiltration test – you may need to seek permission first. The places you select need to be separated as much as possible.

3. The method for measuring infiltration is:

   a. Take a short section of sturdy plastic tubing (drainpipe).

   b. Knock the tubing about 15 cm into the soil.

   c. Pour a set amount of water into the tube and time how long it takes for the water to drain away completely.

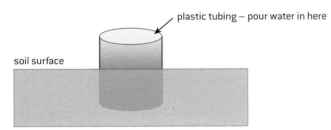

plastic tubing – pour water in here

soil surface

Record your results on a table similar to the one below

	Time taken for 200 ml of water to infiltrate (seconds)					
	Trials					Mean
Surface	1	2	3	4	5	
Grass						
Under trees						
Flower beds						
Footpaths on grass						
Ploughed field						

▲ **Figure 3.47**

You can assume that concrete will not allow any infiltration so please do not try knocking anything into it.

These are just a few possible surfaces you can test. You might test others.

Once you have this data you could measure how much of the study area is covered by each material and extrapolate what may happen if the proportions changed.

Topic 5: Soil systems and terrestrial food production systems and societies

1. Investigate differences in soil profiles (on a slope/around your school)

This could be used to practise the following IA skills:

- Results, analysis and conclusion
- Discussion and evaluation
- Communication

Soils vary enormously and they are easy to investigate. To do a good soil study you really need to see as much of the soil profile as possible. This can sometimes be achieved at natural exposures such as the one shown in figure 3.48. Failing that you will need to expose the soil profile by digging a soil pit. Soil pits are a bit destructive (they are a large hole in the ground) so it is essential to get permission before you start digging.

▲ **Figure 3.48** Naturally exposed soil profile

To investigate soils you should select three or four very different sites, perhaps different locations in your local area. Look at changes with zonation – up a hill, away from a river, away from a road etc. The sites should have undisturbed soil, not a garden or a farm where the soil has been turned over and mixed by humans.

At each site you need to expose the soil profile as much as possible. If you need to dig a soil pit try and remove the vegetation intact as a grass mat. Once you have conducted your fieldwork you can replace the soil in the hole and put the vegetation back on top. Another possible way to check out soil is using a soil auger (figure 3.49). These are screwed into the ground and then pulled straight out. It will bring a column of soil with it which can then be laid out for study.

▲ **Figure 3.49** Soil augers

Method

For each soil profile:

1. Draw a field sketch of the soil profile and identify and label the horizons.

2. Put a stick in the profile at the junction of each horizon then take a picture of the soil profile.

3. There is a range of tests you can do on each horizon of the soil:

   a. colour

   b. texture

   c. pH

   d. water content

   e. organic content.

4. To conduct some of these tests you must take a sample of soil from each horizon and test it back in the lab. This must be done very carefully.

   a. Use ziplock bags that you can write on.

   b. Label each bag with the site and the horizon that the sample is taken from.

   c. Seal the bag tightly to avoid loss of moisture.

5. Once you have finished your tests replace the soil in the soil pit and replace the vegetation mat.

Colour test

The easiest way to do this is to take a small sample of the soil and rub it onto your field sketch. Alternatively there are colour charts (e.g. Munsell colour chart) that can be used as a point of comparison. Match the colour of the soil to the chart and you will get a description and a number for that colour. The advantage of this is that a person in another part of the world could use the chart to see what the colour of the soil in your study is.

Texture

You can take a sample of soil from each horizon, return to the lab and use a sediment settling technique (Mix each soil sample with a known volume of water. Pour each into a large measuring cylinder. Record settling times and then compare depth and types of layers once the soils have settled.). Alternatively you can do field tests to assess the soil texture. You can find a number of 'soil feel tests' – here is one example: http://www.nrcs.usda.gov/wps/portal/nrcs/detail/soils/edu/kthru6/?cid=nrcs142p2_054311

You can print this key (figure 3.50) and use it to assess soil texture.

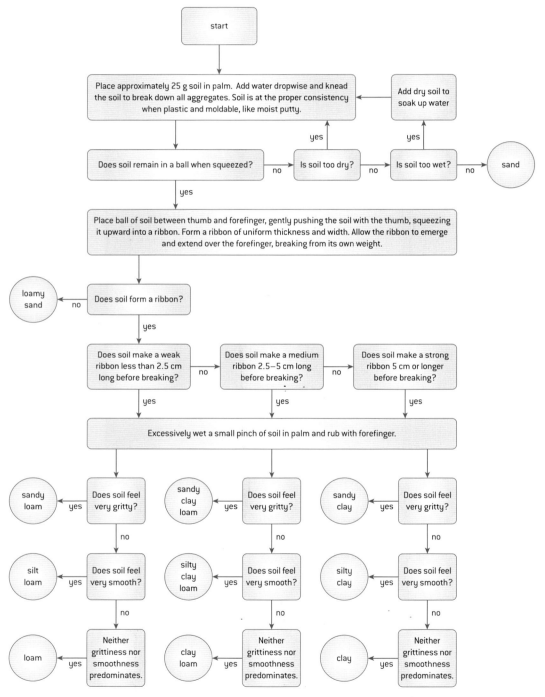

▲ **Figure 3.50** Soil feel test chart. Modified from S J Thien, 1979, 'A flow diagram for teaching texture by feel analysis', *Journal of Agronomic Education*, 8: 54–55

pH

This can be done using soil testing kits back in the lab or in the field.

2. Investigate soil erosion

In this investigation you can look at the impact of vegetation cover on soil erosion.

> This could be used to practise the following IA skills:
> * Identifying context
> * Results, analysis and conclusion
> * Discussion and evaluation

Method

1. Take four large soda bottles.

2. Cut them as shown in figure 3.51.

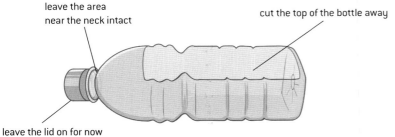

leave the area near the neck intact

cut the top of the bottle away

leave the lid on for now

▲ **Figure 3.51** Cut out bottle for soil experiment

3. Place soil in all of the bottles, as follows:

   a. Bottle 1: Leave with just soil.

   b. Bottle 2: Cover with a mulch or leaf litter.

   c. Bottle 3: Plant half with grass seeds.

   d. Bottle 4: Plant completely with grass seeds.

4. Place bottles 3 and 4 in a sunny position, water regularly (not too much), and leave for a while until the grass has grown.

5. Get the four bottles ready.

   a. Take bottle 1 and place it on a table with the lid on hanging over the edge.

   b. Place a measuring cylinder under the bottle lid (leave the lid on for now).

   c. Fill a watering can with a set amount of water. The amount will depend on how big the soda bottles are – use your judgment.

   d. Remove the lid and water the bottle with the prepared water in the watering can. Pour at a constant rate and for a set amount of time.

   e. Leave it in place for about 30 minutes.

6. Repeat with bottles 2, 3 and 4.

7. After about 30 minutes check the measuring cylinders and see how much soil is in each of them.

8. For statistically valid results you should repeat this 5 times for each cover type.

The same experiment can be done to assess the impact of slope angle on soil erosion.

Instead of different contents in each of the bottles, set them all up with the same 'filling' but when you water them place them at different angles.

3. Investigate the effect of soil salinization on plant growth (or seed germination)

This could be used to practise the following IA skills:

- Identifying context
- Results, analysis and conclusion
- Discussion and evaluation
- Application
- Communication

You could investigate:

- the impact of different salt concentrations on the growth of a particular plant (germination of particular seeds)
- the impact of salinity on different types of plants (seeds).

4. Investigate food consumption and/or production patterns

This could be used to practise the following IA skills:

- Identifying context
- Results, analysis and conclusion
- Discussion and evaluation
- Application
- Communication

There is a range of data available on food consumption and production. You could look at:

- The relationship between food consumption and level of development of selected countries. Good websites for food consumption:
  - http://chartsbin.com/view/1150 (daily calorie intake / capita)
  - http://en.wikipedia.org/wiki/List_of_countries_by_food_energy_intake (daily kilocalorie and kilojoule intake / capita)
  - http://www.gapminder.org/data/ (daily kilocalorie intake / capita)
  - https://www.cia.gov/library/publications/the-world-factbook/
    - Select country of interest.
    - Select 'people and society'.
    - Scroll down to find 'obesity' %adults.
- The relationship between food production and level of development of selected countries. Good websites for food production
  - http://www.indexmundi.com/facts/indicators/AG.PRD.FOOD.XD/rankings

- ■ Nationmaster.com has a range of agricultural data.
  - ○ Go to http://www.nationmaster.com/
  - ○ Under categories select 'agriculture' and there are a number of production indicators given.
- ● The relationship between consumption and production.

Method

This investigation is looking at the relationship between the level of development of a country and its food production/consumption.

1. Choose a sampling system to use. Systematic sampling – every nth country from a ranked list.

   Random stratified sampling – taking a set number of countries from each of the development categories:

   a. very high human development

   b. high human development

   c. medium human development

   d. low human development.

2. Decide which method you will use then go to the HDI list http://hdr.undp.org/en/data:

   a. Click on Human Development Index (HDI) value.

   b. This will give a list of all countries according to the categories above.

   c. Use the sampling method you have selected to generate a minimum of 30 countries and record them on a data recording table (similar to that below).

Country	Daily calorie intake per capita (Chartsbin)	Food production index (baseline 100) (Indexmundi)
Norway	3,460	95.85
New Zealand	3,150	106.60

▲ **Figure 3.52**

3. Go to http://chartsbin.com/view/1150 (or whatever website you select). Hover over each country on your list and record the daily calorie intake per capita.

4. Go to Indexmundi http://www.indexmundi.com/

5. Find each country on your list and record the food production index.

6. Draw a scattergraph.

7. Conduct an appropriate statistical test.

5. Investigate your food consumption

This could be used to practise the following IA skills:

- Results, analysis and conclusion
- Discussion and evaluation

There are many aspects of your food consumption that you could investigate. Here are two fun activities. They are interlinked and follow on from each other.

Method

1. Select 30 food items in your home and record the country they are from. Try and go for a mix of items.

    a. For packaged food: check the labels and find country of origin.

    b. For fruit and vegetables and meat and fish: you may need to go to where the items were purchased.

2. Go to http://www.foodmiles.com/. Follow the instructions on the right-hand side of the page and record the:

    a. food miles

    b. kg CO_2 or kg Carbon if the distance was covered by plane

    c. kg CO_2 or kg Carbon if the distance was covered by car

    d. kg CO_2 or kg Carbon if the distance was covered by train.

| | | | Kg CO_2 or kg Carbon | | |
Food item	Country of origin	Food miles	Plane	Car	Train
Tomato sauce	USA				
Cereal	UK				
Olives	France				

▲ **Figure 3.53**

3. The clearest way to present the data on the country of origin is a flow line map. See http://geographer-at-large.blogspot.co.uk/2011/11/map-of-week-11-28-2011the-spread-of.html for an example of a flow line map by Haisam Hussein on the spread of disease.

4. You can then calculate the amount of carbon generated by getting your food items to you and present it in the clearest way possible.

BEWARE

Having done all this, check out this website: http://shrinkthatfootprint.com/food-miles. What conclusions can you draw?

Topic 6: Atmospheric systems and societies

1. Investigate acid deposition in a range of countries

This could be used to practise the following IA skills:

- Identifying context
- Results, analysis and conclusion
- Discussion and evaluation
- Application

This investigation could look at:

- The percentage of the land that has acidification problems. Go to the Nationmaster website: http://www.nationmaster.com/country-info/stats/Environment/Acidification
- Sulphur emissions either per person or total emissions:
  - Go to Gapminder: http://www.gapminder.org
  - Click on data tab.
  - In the 'search' box enter 'sulphur'.
  - Download the Excel file with the data you want – per person or total emissions.

The method for this investigation is similar to that for investigating the relationship between food and development level (Topic 5). This will give you the selection of countries and then you can collect the data on sulphur emissions.

2. Investigate air pollution using a biotic indicator (lichens)

This could be used to practise the following IA skills:

- Identifying context
- Results, analysis and conclusion
- Discussion and evaluation
- Application
- Communication

A common source of air pollution is the combustion of fossil fuels so for this investigation we are using the busy road. Thermal power stations also burn fossil fuels but the distances over which the pollution can travel are very large.

Lichens grow in exposed places such as on rocks and tree bark. They absorb water and nutrients from rain water so if the rain is polluted the lichens can be damaged. You will have to look up the lichens common in your area, but as a general rule:

- crusty lichens are the most tolerant of pollution
- leafy lichens tolerate a little air pollution
- bushy lichens are intolerant of any air pollution and will not be present in polluted air.

Crusty lichens   Leafy lichens   Bushy lichens

▲ **Figure 3.54** Types of lichens

Method

1. Research your local area to find out which lichens are commonest.

2. Find a suitable study site with an environmental gradient from polluted (e.g. roadside) to less polluted (e.g. woodland or forest).

3. Lay out a minimum of three transects starting as close to the road as you can and moving into the forest.

4. Select a relevant sampling interval – this will depend on how far into the forest you go but every 10 m for 100 m is usually suitable.

5. At each point select the three trees closest to the transect line.

    a. On the side of the tree closest to the road measure a set distance (1 m) from the base of the tree.

    b. Using a suitably sized quadrat (10 cm × 10 cm) use a suitable method to assess the abundance of each species of lichen (see page 66).

6. Repeat this along all three transects.

7. Throughout the investigation make sure all the trees are the same species.

3. Investigate the impact of the albedo of different surfaces on the surrounding air temperature

The albedo is the proportion of sunlight reflected back from an object.

- Light colours have much higher albedo rates than dark colours.
- Artificial surfaces have much higher rates than natural ones.

You can look up the actual differences for yourself.

Method

1. Select a range of surfaces – have a mix of natural and manmade, for example:

    a. grass or other vegetation

    b. glass

   c. mirrors

   d. bare soil

   e. snow or ice

   f. water

   g. concrete.

2. The easiest way to collect the data is with a data logger and temperature sensor. This must be set up in a safe area away from areas where it could be interfered with.

   a. Set up the data logger and temperature sensor at a fixed height above the selected surface.

   b. Record the temperature changes during sunlight hours.

   c. Repeat for all surfaces you have selected.

3. If you do not have access to a data logger:

   a. Set up an analogue thermometer at a fixed height above the selected surface.

   b. Check the temperature every hour during sunlight hours.

   c. Repeat for all surfaces you have selected.

4. You can then compare the differences in temperature and extrapolate what impact this may have on the global climate.

Topic 7: Climate change and energy production

1. Investigate the impact of different gases on the temperature of a body of air

In Topic 7 greenhouse gases (GHG) and the global warming potential (GWP) are discussed. This is something you could put to the test.

Method

1. Set up identical soda bottles with different gases in.

    a. Normal air

    b. CO_2 in varying ppm

    c. Humid air

    d. Dry air

2. Place a thermometer in each of the bottles and seal them.

3. Check the temperature in the bottles regularly.

This investigation could also be done with coloured filters around the bottles to see what the impact of colour is on heating.

Topic 8: Human systems and resource use

1. Graphically represent changes in human population size over time

This could be used to practise the following IA skill:

- Results, analysis and conclusion

1a. Draw a line graph to show world population change from 10,000 years BCE (before common era) to 2050

Year	Global population (millions)
−10000 BCE	5
−8000	7
−5000	10
−4000	7
−3000	14
−2000	27
−1000	50
−500	100
−400	162
−200	191
0 CE	242
200	223
400	148
600	693
800	512
1000	277
1200	398
1400	362
1600	545
1800	929
1900	1615
2000	6089
2010	6866
2015	7253
2020	7702
2025	7952
2030	8236
2035	8528
2040	8774
2045	8992
2050	9234

▲ **Figure 3.55** Statistics are taken from http://en.wikipedia.org/wiki/World_population_estimates

1b. Draw a line graph to show the population growth of your own country

Method

- Go to http://www.populstat.info/
- Click on the appropriate continent.
- Click on the letter for the country you are interested in.
- Click on 'of the whole country'.
- That will give you demographic data for the whole country for a number of years.

2. Draw a population pyramid using Excel (or similar)

This could be used to practise the following IA skill:

- Results, analysis and conclusion

There are a number of websites that take you through the process of drawing a population pyramid using Excel – here are a few:

- http://www.excel-exercise.com/charts/population-pyramid/
- http://www.uvm.edu/~agri99/spring2004/Population_Pyramids_in_Excel.html
- http://www.abs.gov.au/websitedbs/CaSHome.nsf/Home/GeoQ06B+Draw+a+population+pyramid+using+Excel

3. Draw line graphs to show changes in birth rates, death rates and growth rates over time for a country of your choice

This could be used to practise the following IA skill:

- Results, analysis and conclusion

Year	Crude birth rate (births/1000/year)	Crude death rate (deaths/1000/year)
1910	14.7	30.1
1915	13.2	29.5
1920	13.0	27.7
1925	11.7	25.1
1930	11.3	21.3
1935	10.9	18.7
1940	10.8	19.4
1945	10.6	20.4
1950	9.6	24.1
1955	9.3	25.0

Tip

Make sure the interval between years is the same.

1960	9.5	23.7
1965	9.4	19.4
1970	9.5	18.4
1975	8.8	14.8
1980	8.7	15.9
1985	8.7	15.8
1990	8.6	16.7
1995	8.8	14.8
2000	8.7	14.7
2005	8.2	14.0

(Source: http://www.infoplease.com/)

▲ **Figure 3.56** Fertility and mortality rates in the USA 1910–2005

You could calculate growth rate and draw a compound line graph for the USA.

There are plenty of websites with demographic information:

- http://www.infoplease.com/ – use the search engine on the site to find what you want.

- http://www.gapminder.org/
  - Go to Gapminder world tab and you can change the axes of the graph to find statistics.
  - Go to data tab and you can select the data you want using the search box.

- https://www.cia.gov/library/publications/the-world-factbook/ – go to the country of your choice and find the statistics you need.

- http://www.indexmundi.com/ – select the area or the statistics you want.

There is a wide range of investigations you can do in this topic.

4. Investigate the relationship between level of development and demographics

This could be used to practise the following IA skills:
- Planning
- Results, analysis and conclusion
- Communication

You can combine any of the following:

Level of development as measured by	Demographic factor
HDI: rank or actual number	**Fertility rates:** CBR, total fertility rate
GDPUS$: per capita, whole country	
Education: years in school, literacy rate, proportion of males:females	**Mortality rates:** life expectancy, CDR, IMR
Health: hospital beds/person, doctors/person	**Growth rate:** as a %, doubling time

▲ **Figure 3.57**

5. Investigate environmental value systems (EVS) and resource exploitation

This could be used to practise the following IA skills:

- Identifying context
- Results, analysis and conclusion
- Discussion and evaluation

Imagine you are presented with a research question asking you to investigate environmental attitudes towards resource exploitation. This could be linked to age, gender or education (see section 3c. Questionnaires for details and a reminder of the three main EVSs).

The method is the same as the EVS investigation under Topic 4.

Have this table on a separate sheet for respondents to view with the photos.

I am technocentric and I believe whatever problems we cause, we can solve them.	I am ecocentric and I believe we need the Earth more than it needs us.
- We are the Earth's most important species, we are in charge. - There will always be more resources to exploit. - We will control and manage these resources and be successful. - We can solve any pollution problem that we cause. - Economic growth is a good thing and we can always keep the economy growing.	- The Earth is here for all species. - Resources are limited. - We should manage growth so that only beneficial forms occur. - We must work with the Earth, not against it.

We exploit a wide range of the Earth's resources. For each of the photographs please indicate which response you agree with.

Overfishing		Deforestation	
	I am technocentric on this issue		I am technocentric on this issue
	I am ecocentric on this issue		I am ecocentric on this issue
Opencast mining		Soil degradation through overcropping	

	I am technocentric on this issue			I am technocentric on this issue
	I am ecocentric on this issue			I am ecocentric on this issue
Fur trade		Whaling		
	I am technocentric on this issue			I am technocentric on this issue
	I am ecocentric on this issue			I am ecocentric on this issue
Ruined landscapes		Population growth causing loss of habitat		
	I am technocentric on this issue			I am technocentric on this issue
	I am ecocentric on this issue			I am ecocentric on this issue

There is a wide range of photos you could use – make it relevant to where you live.

6. Investigate solid domestic waste generated in school compared to your home

This could be used to practise the following IA skills:

- Results, analysis and conclusion
- Discussion and evaluation
- Application
- Communication

Make a list of all the solid domestic waste you think you produce in your home and in school. It is best to give categories of waste (see table below).

Quantity of waste (kg)								
Week	1		2		3		4	
Category of waste	Home	School	Home	School	Home	School	Home	School
Paper								
Plastic (all sorts)								
Metal								
Food leftovers								
Glass								
Batteries								

To make data easy for you to collect:

1. Set up bags or boxes clearly labelled with each waste category.

2. Ask your family to place their waste into the different labelled containers.

3. School may have fewer categories of waste so you could start a recycling system at school and this would separate out the types of waste for you.

7. Investigate attitudes towards the intrinsic value of landscapes

This could be used to practise the following IA skills:

- Planning
- Results, analysis and conclusion

This is a difficult investigation to do in a generalized way so this is very specific to the individual – who we are and our backgrounds, cultural, social and economic. All these strongly affect how we value a landscape so you will have to make some adjustments to suit your audience.

Method

1. Do a general opinion poll about what people value in a landscape – maybe use some photo stimuli like the ones on the next page. Pick pictures that you really like and ask for people's opinions about each one. Here are some possible responses to this question.

It is:

a. peaceful/tranquil/quiet

b. beautiful

c. wild/free

d. open

e. natural

f. spiritual

g. colourful

h. a good source of income.

2. Now select a different set of pictures from close to where you live.

3. For a minimum of 30 respondents ask them:

   a. to rank the pictures from best to worst

   b. if they feel it is acceptable to develop that area for a theme park (or something relevant to the area)

   c. how they rate the landscape.

The area looks	4	3	2	1	0
Peaceful/tranquil/quiet					
Beautiful					
Wild/free					
Open					
Natural					
Spiritual					
Colourful					

8. Simulation on running your own country

This is a good way to see what impact decisions can have on the human population and the environment.

This could be used to practise the following IA skills:

- Results, analysis and conclusion
- Discussion and evaluation

Method

1. Go to http://www.catchmentdetox.net.au/

2. Read the notes that start 'what is a catchment' (1–9).

3. You can also read 'How to play' and run the tutorial.

4. When you are ready click 'play game'.

5. Read 'Catchment challenge'.

   a. Set and record your levels under catchment manager.

   b. Take action in the catchment and record them.

c. Click 'next turn' (bottom right corner).

d. Check your mail – find the icon below on the left of the page and record any advice or comments.

e. Check and record your water statistics:

    i. water status

    ii. rainfall

    iii. water for production.

f. Check and record your game statistics:

    i. environmental health

    ii. water for the environment

    iii. water quality

    iv. catchment biodiversity

    v. available cash

    vi. food production – if you do not do this your people die!!

Year 1							
Catchment manager	LEVEL	Actions taken		Water statistics		Game statistics	
Water restrictions		Agriculture		Water status		Environmental health	
Water solving technology		Industry and tourism		Rainfall		Water for the environment	
Environmental research		Irrigation and dams		Water for production		Water quality	
		Eco				Catchment biodiversity	
Mail messages?						Available cash	
						Food production	

Year 2							
Catchment manager	LEVEL	Actions taken		Water statistics		Game statistics	
Water restrictions		Agriculture		Water status		Environmental health	
Water solving technology		Industry and tourism		Rainfall		Water for the environment	
Environmental research		Irrigation and dams		Water for production		Water quality	
		Eco				Catchment biodiversity	
Mail messages?						Available cash	
						Food production	

▲ **Figure 3.58** Data collection table for 'Catchment detox'

You can run the country for as long as you like (up to a maximum of 100 turns).

There are many guides to revising for exams; this is not one of them. Everyone finds a slightly different way to revise but there are tips that are based on research into the best ways of remembering facts and recalling information under exam conditions. Here are some of them.

Tips for revising

1. Start earlier than you think

▲ **Figure 4.1** Plan ahead – don't get a shock when you revise

Last minute cramming does not work. Your brain needs time to absorb information so start months before the exams, not weeks or days.

Revision timetables do work but don't spend all your time writing the timetable or revising it if you fall behind (as Rimmer did in Red Dwarf, see page 138!).

Revision means revisiting something you have already covered to refresh your understanding and knowledge of the topic and maybe to expand these. It should not be new to you at this point as you should already have covered the course.

Evidence is that if you have a longer gap between revision sessions, you score higher. The longer the gap, the better you do.

Less is more but only if you spread it out and revisit topics.

'Never put off till tomorrow what may be done the day after tomorrow just as well.' Mark Twain

Do it now. Don't put it off (procrastinate). The exams will not be postponed because you need another week of revision. Rearranging your desk, sharpening pencils, writing at last to your old great aunt are all good things to do but do them at the right time – probably after the last exam.

2. Keep healthy

▲ **Figure 4.2** Stay healthy and be awake at the right times

You cannot revise effectively for 16 hours a day. Work out if you are a morning or evening person. In this case, that means whether you work better earlier or later in the day, not when you like to play. Plan your revision work for the time of day when you are most receptive to it – usually not 3 am the night before the exam.

Reward yourself with breaks every hour, not every five minutes. Get up, walk, run, be active. Do something else for a short time.

Sleep and eat properly.

Don't hang out with your exam-tense colleagues. You probably won't help them much and might end up harming yourself. Be a bit selfish at this time.

3. Use tricks that help you

If it is index cards in your jeans pocket, use that. If it's post-it notes all over your house or on the fridge, use these. If you learn best when you are jumping up and down or pacing your room with music on, do that. Work out what type of learner you are. Many of us are visual learners so we need to see it. Others remember better if hearing it or moving around at the same time.

Research has shown that only reading something is not a very effective way to remember it. You need to do something – write, draw, annotate, condense your notes – in order to recall successfully.

Remember that only you see your revision notes so they do not have to be perfect. These notes do not need to be a work of art or in wonderful handwriting. As long as you understand them, that is fine.

Use diagrams, mind maps, doodles, whatever you like, but the action of actively writing something down is far, far better than just passively reading something. Limit yourself to three colours if you need colour.

4. Practise, practise, practise

You are probably reading this because you want to practise before your exams and that is a good thing. Practise writing by hand for long periods if you need to. Practise answering past exam papers. Practise writing short essays. Practise answering a paper in the time allowed for it. You can learn the **skill** of sitting an exam.

5. Be there

▲ **Figure 4.3** Get to the exams in good time – don't miss the bus!

You could have revised brilliantly, know it all, finally understand that concept, but if you miss the exam, you miss the exam. Check and recheck the exam timetable. Be there in good time, don't rush. Make sure you have the correct equipment – spare pens, pencils, ruler, calculator, water, sweets if allowed etc. Stay to the end of the exam. Don't walk out early. Try all the questions even if you are guessing. There is no negative marking in ESS.

Explaining ESS exams

By the time you sit the IB Diploma exams, you will have completed all the coursework or internal assessment. While you won't know what marks you were awarded for the IA until you get your total IB score, you should be content to know that you have some marks, maybe up to two grades' worth, already.

The exams are called external assessment and in ESS consist of two papers.

Paper 1 is based on a case study in a Resource Booklet (RB) with data relating to one location, ecosystem or ESS theme. It is one hour long and counts for 25% of the final grade.

You need to demonstrate the ability to apply your knowledge from the course to novel situations in a holistic way.

You evaluate information, explain your viewpoints and the views of others, and justify knowledge claims.

The last question on the paper is a more holistic one, in which you have to answer based on all the information provided in the case study.

Paper 2 is a two-part exam and is two hours long. It counts for 50% of the final grade.

Section A is worth 25 marks and has 3–5 short answer and data-based questions. The marks for the sub-questions vary from 1–3 marks. The idea of this section is to make sure all the topics are covered – so you need to revise everything. It should take you 40 minutes.

Section B gives you four essay questions and you should answer two. Each essay has parts (a), (b) and (c) and each essay should take you 40 minutes. The final part of each essay, part (c), is worth 9 marks out of the 20 available for each essay, is likely to be open-ended, and expects you to be holistic in your answer – that means use all your ESS knowledge to answer it. It may be related to the 'big questions' of the ESS guide. Part (c) is marked using generic markbands along with specific suggested responses in a markscheme. If you want the top marks here you are going to have to look for the links between environment and society.

What you need to do

The table below summarizes the details of each paper given above as well as showing you what objectives are assessed on each paper.

External assessment	Duration	Assessment objectives	What is in the paper	Total marks	Weighting (%)
Paper 1	1 hour	Objectives 1–3	Case study	35	25
Paper 2	2 hours	Objectives 1–3	Section A: short-answer questions Section B: two essays from a choice of four	65 Section A: 25 marks Essays: 20 marks each	50

▲ **Figure 4.4** Times, marks and content of ESS papers

The internal assessment takes 10 hours and covers assessment objectives 1–4. It is scored out of 30 marks and is worth 25% of the final marks.

The **IB assessment objectives** tell you what will be assessed and this is what the IB wants you to achieve.

1. Demonstrate knowledge and understanding of relevant:
    - facts and concepts
    - methodologies and techniques
    - values and attitudes.

2. Apply this knowledge and understanding in the analysis of:
    - explanations, concepts and theories
    - data and models
    - case studies in unfamiliar contexts
    - arguments and value systems.

3. Evaluate, justify and synthesize, as appropriate:
    - explanations, theories and models
    - arguments and proposed solutions
    - methods of fieldwork and investigation
    - cultural viewpoints and value systems.

4. Engage with investigations of environmental and societal issues at the local and global level through:
    - evaluating the political, economic and social context of the issues
    - selecting and applying the appropriate research and practical skills necessary to carry out investigations
    - suggesting collaborative and innovative solutions that show awareness and respect for the cultural differences and value systems of others.

Read the course guide

Make sure you have your own copy of the course guide or that your school gives you access to it.

The guide tells you what the course covers, but more importantly it gives you a great deal of guidance and advice. So, make sure you read the guide carefully and look at these sections:

1. Nature of the subject, aims, assessment objectives – these tell you what the course covers, how to approach it and what we hope you will be able to do when you have completed it.

2. Check the topics and sub-topics. Each is arranged like this:

1.1 Sub-topic		
Significant ideas: Describes the overarching principles and concepts of the sub-topic.		
Main ideas	**Knowledge and understanding:** • This section will provide specifics of the content requirements for each sub-topic. **Applications and skills:** • This section gives details of how you can apply your understanding. For example, these applications could involve discussions of viewpoints or evaluating issues and impacts.	**Guidance:** ◄ • This section will provide specifics and give constraints to the requirements for the understandings and applications and skills. **International-mindedness:** ◄ • Examples of ideas that teachers can easily integrate into the delivery of their lessons. **Theory of knowledge:** ◄ • Examples of TOK knowledge questions. **Connections**: syllabus and cross-curricular links.	Tells you what depth is required International-mindedness and TOK ideas – see reflective background questions

Left-hand column can all be assessed.

▲ **Figure 4.5** ESS course guide layout

Other things you should read

Past papers, specimen papers and markschemes are useful, though be careful because the syllabus for 2015 onwards has been changed and some of the topics have moved, changed and been updated. Past papers and markschemes are published by the IB and your teacher should have some. Towards the end of the course, you should look at and do some of these. Practise getting your timing right so that you do not run out of time in the exam. Read the markschemes carefully so you can see how the examiners award marks. Remember that ESS requires you to apply your knowledge to new situation, so you can't rely on simply learning the 'correct' answers.

Subject reports are written by the senior examiner after every exam session. They report on each question and how it was answered as well as commenting on the IA. They are really useful as there is advice for future candidates in every one and each question is analysed for the answers given and strengths and weaknesses that candidates showed.

You can get copies of subject reports from your teacher.

Know the command terms

The IB assesses how well you do in the assessment objectives using **command terms**.

Command terms are action verbs that tell you what to do in the exams and it is very important that you can recognize them, know what they mean and do what they ask you to do.

Look at Figure 4.6 and you will see that what you need to do gets harder as you go down the list. Now compare the command terms to the assessment objectives.

We cannot emphasize enough how important it is for you to know these command terms.

These terms tell you how to answer a question and in what depth. Sometimes other terms may be used in the exam questions but knowing these command terms will help you greatly.

They are classified into three groups based on the objectives.

Objective 1	
Define	Give the precise meaning of a word, phrase, concept or physical quantity.
Draw	Represent by means of a labelled, accurate diagram or graph, using a pencil. A ruler (straight edge) should be used for straight lines. Diagrams should be drawn to scale. Graphs should have points correctly plotted (if appropriate) and joined in a straight line or smooth curve.
Label	Add labels to a diagram.
List	Give a sequence of brief answers with no explanation.
Measure	Obtain a value for a quantity.
State	Give a specific name, value or other brief answer without explanation or calculation.
Objective 2	
Annotate	Add brief notes to a diagram or graph.
Apply	Use an idea, equation, principle, theory or law in relation to a given problem or issue.
Calculate	Obtain a numerical answer showing the relevant stages of working.
Describe	Give a detailed account.
Distinguish	Make clear the differences between two or more concepts or items.
Estimate	Obtain an approximate value.
Identify	Provide an answer from a number of possibilities.
Interpret	Use knowledge and understanding to recognize trends and draw conclusions from given information.
Outline	Give a brief account or summary.
Objectives 3 and 4	
Analyse	Break down in order to bring out the essential elements or structure.
Comment	Give a judgment based on a given statement or result of a calculation.
Compare and contrast	Give an account of similarities and differences between two (or more) items or situations, referring to both (all) of them throughout.
Construct	Display information in a diagrammatic or logical form.
Deduce	Reach a conclusion from the information given.

Demonstrate	Make clear by reasoning or evidence, illustrating with examples or practical application.
Derive	Manipulate a mathematical relationship to give a new equation or relationship.
Design	Produce a plan, simulation or model.
Determine	Obtain the only possible answer.
Discuss	Offer a considered and balanced review that includes a range of arguments, factors or hypotheses. Opinions or conclusions should be presented clearly and supported by appropriate evidence.
Evaluate	Make an appraisal by weighing up the strengths and limitations.
Explain	Give a detailed account, including reasons or causes.
Examine	Consider an argument or concept in a way that uncovers the assumptions and interrelationships of the issue.
Justify	Provide evidence to support or defend a choice, decision, strategy or course of action.
Predict	Give an expected result.
Sketch	Represent by means of a diagram or graph (labelled as appropriate). The sketch should give a general idea of the required shape or relationship, and should include relevant features.
Suggest	Propose a solution, hypothesis or other possible answer.
To what extent	Consider the merits or otherwise of an argument or concept. Opinions and conclusions should be presented clearly and supported with appropriate evidence and sound argument.

▲ **Figure 4.6** Command terms and how they relate to IB objectives

Exam strategies

i. Thinking like an examiner

Imagine you are an IB examiner instead of an IB candidate. What is it like at the other end? Examiners have been approved and trained by the IB and they are subject specialists. They undergo training in marking and the marking of each examiner is rigorously monitored. Many are also IB teachers. Some of your teachers may be IB examiners. Ask them what it is like.

There is a three-week window after you sit the exam in which all the marking should be done. E-marking in the IB Diploma means you write your answers on a script by hand and then the scripts are sent securely by your school to a scanning centre where they are all scanned. Examiners see individual scripts on their computer screens. The scripts are allocated at random so a single examiner does not mark the scripts for your whole class but a random selection from any candidate anywhere in the world.

▲ **Figure 4.7** Marking IB exams

▲ **Figure 4.8** Sitting IB exams

The examiners who mark your papers are human beings too. They do their very best to mark accurately according to the markscheme, to find marks to award to you in your scripts and to decipher your handwriting if it is not too clear. Often more than one examiner sees your script and checks the marking and sometimes up to four examiners could mark your script.

But you can help them to mark your work by making it easier for them.

Is your handwriting clear enough? Most of us do not write that much but in the IB exam season, you could be writing for up to six hours a day.

Can you do this? Practise using a pen when you revise and ask someone to check your handwriting for legibility and clarity. You are not allowed to write in pencil so make sure you practise in pen.

ii How IB assessment works

There are thousands of IB examiners all around the world. Each subject has a chief and several deputy chief examiners and marking teams made up of dozens of examiners depending on subject size.

The IB Diploma is assessed using criterion-based assessment. That means that your achievement levels are measured relative to the criteria, not relative to the other candidates in that exam. In theory, that means everyone could gain a grade 7 or a grade 1, but in practice most of us are somewhere in the middle with a few at each end.

This is why we cannot say a mark of say 78 out of 100 is a grade 7. This varies slightly from year to year as the papers can never be exactly comparable in terms of their difficulty from year to year. The criterion-based referencing means it is fairer to you.

To give you some idea of grade distributions, figure 4.9 shows the last three exam sessions of published results.

Tip

Exam tips

- Check that you can write legibly for long periods of time.
- If you need to go over the space allocated for a response, make sure you show clearly where the continuation is written.
- Only write inside the boxes for your responses.
- Diagrams can help.
- Basics – check that you have spare pens, pencils, a ruler and a calculator with a new battery in it.

ESS grade distributions				Percentage of candidates gaining each grade							
	Level	Candidates	Mean grade	Grade 1	Grade 2	Grade 3	Grade 4	Grade 5	Grade 6	Grade 7	Total
May 2014	SL	9520	4.2	0	9	21	28	27	21	10	100
Nov 2014	SL	730	4.3	1	6	24	28	20	14	7	100
May 2013	SL	8944	4.22	0	9	21	29	25	13	3	100

▲ **Figure 4.9** ESS grade distributions in May 2013, Nov 2014 and May 2014

So how do examiners set the grade boundaries?

- There is a set amount of total marks available for each paper.
- There is a set total amount of marks available for the IA.
- For Paper 1 and Paper 2, the senior examiners then determine the grade boundary marks between grades 3/4, 6/7 and 2/3 by using their experience of the responses, professional judgment, written grade descriptors, statistical comparisons and expectations of experienced teachers (which is where your predicted grades come in). A great deal of thought, care, discussion and evidence goes into this as all are very aware that it is so important to get it right.
- IA grade boundaries do not alter from year to year and are published.
- The resulting marks are weighted accordingly: Paper 1 25%; Paper 2 50%; IA 25%.
- The marks and boundaries are then rounded to the nearest whole number. 0.5 is rounded up.

This all means that your marks do not depend on how your classmates or the other candidates have done but on how well you did in the tests.

What are the grade descriptors?

Grade descriptors are written statements for each grade and describe how a student getting that grade has performed.

The first paragraph describes how you performed in the exam paper. The second describes how you performed in practical work. You will see when you read them that the tasks are the same; it is how you performed in it that changes.

Grade descriptors

Grade 7

Demonstrates: comprehensive understanding of factual information, terminology, concepts, methodology and skills; high level of ability to apply and synthesize information, concepts and ideas; quantitative and/ or qualitative analysis and thorough evaluation of data; fully reasoned, balanced judgments based on holistic consideration of multiple viewpoints, methodologies or opinions; communication of a well-reasoned and justified personal viewpoint, while critically appreciating the views, perceptions and cultural differences of others; a high level of proficiency in communicating logically and concisely using appropriate terminology and conventions; insight and originality. ← Exam paper

Demonstrates personal skills, perseverance and responsibility in a wide variety of investigative activities in a very consistent manner. Works very well within a team and approaches investigations in an ethical manner, paying full attention to environmental impact. Displays competence in a wide range of investigative techniques, paying considerable attention to safety, and is fully capable of working independently. ← Internal assessment

Grade 6

Demonstrates: broad detailed understanding of factual information, terminology, concepts, methodology and skills; good level of ability to apply and synthesize information, concepts and ideas; quantitative and/or qualitative analysis and evaluation of data; reasoned, balanced judgments based on holistic consideration of multiple viewpoints, methodologies or opinions; communication of a reasoned and justified personal viewpoint, while critically appreciating the views or perceptions and cultural differences of others; a competent level of proficiency in communicating logically and concisely using appropriate terminology and conventions; occasional insight and originality. ← Exam paper

Demonstrates personal skills, perseverance and responsibility in a wide variety of investigative activities in a very consistent manner. Works well within a team and approaches investigations in an ethical manner, paying due attention to environmental impact. Displays competence in a wide range of investigative techniques, paying due attention to safety, and is generally capable of working independently. ← Internal assessment

Grade 5

Demonstrates: sound understanding of factual information, terminology, concepts, methodology and skills; competent level of ability to apply and synthesize information, concepts and ideas; some quantitative and/or qualitative analysis and evaluation of data; balanced judgments based on holistic consideration of multiple viewpoints, methodologies or opinions; communication of a reasoned personal viewpoint, while appreciating the views or perceptions and cultural differences of others; a basic level of proficiency in communicating logically and concisely using appropriate terminology and conventions.

`[Exam paper]`

Demonstrates personal skills, perseverance and responsibility in a variety of investigative activities in a fairly consistent manner. Generally works well within a team and approaches investigations in an ethical manner, paying attention to environmental impact. Displays competence in a range of investigative techniques, paying attention to safety, and is sometimes capable of working independently.

`[Internal assessment]`

Grade 4

Demonstrates: secure understanding of factual information, terminology, concepts, methodology and skills with some information gaps; some level of ability to apply and synthesize information, concepts and ideas; descriptive ability with quantitative and qualitative data, limited analytical skills and rudimentary evaluation of some contrasting viewpoints or methodologies or opinions; communication of a personal viewpoint, some awareness of perceptions and cultural differences of others; a basic limited proficiency in communicating using appropriate terminology and conventions.

`[Exam paper]`

Demonstrates personal skills, perseverance and responsibility in a variety of investigative activities, although displays some inconsistency. Works within a team and generally approaches investigations in an ethical manner, with some attention to environmental impact. Displays competence in a range of investigative techniques, paying some attention to safety, although requiring some close supervision.

`[Internal assessment]`

Grade 3

Demonstrates: limited understanding of factual information, terminology, concepts, methodology and skills with information gaps; basic level of ability to apply and synthesize information, concepts or ideas; descriptive ability with quantitative and qualitative data; vague consideration of some contrasting viewpoints or methodologies or opinions; basic communication of a personal viewpoint, weak awareness of perceptions and cultural differences of others; communication lacking in clarity, with some repetition or irrelevant material.

`[Exam paper]`

Demonstrates personal skills, perseverance and responsibility in some investigative activities in an inconsistent manner. Works within a team and sometimes approaches investigations in an ethical manner, with

`[Internal assessment]`

some attention to environmental impact. Displays competence in some investigative techniques, occasionally paying attention to safety, and requires close supervision.

Grade 2

Demonstrates: partial understanding of factual information, terminology, concepts, methodology and skills with some fundamental information gaps; little ability to apply and synthesize information, concepts or ideas; limited descriptive ability with quantitative and qualitative data; no consideration of contrasting viewpoints or methodologies or opinions; weak communication of a personal viewpoint, occasional awareness of perceptions and cultural differences of others; communication frequently lacking in clarity, repetitive, irrelevant or incomplete.

← Exam paper

Rarely demonstrates personal skills, perseverance or responsibility in investigative activities. Works within a team occasionally but makes little or no contribution. Occasionally approaches investigations in an ethical manner, but shows very little awareness of the environmental impact. Displays competence in a very limited range of investigative techniques, showing little awareness of safety factors and needing continual and close supervision.

← Internal assessment

Grade 1

Demonstrates: very little understanding of factual information, terminology, concepts, methodology and skills with many fundamental information gaps; inability to apply and synthesize information, concepts or ideas; inability with handling quantitative and qualitative data; no consideration of contrasting viewpoints or methodologies or opinions; lacks communication of a personal viewpoint, or awareness of perceptions and cultural differences of others; communication mostly lacking in clarity, repetitive, irrelevant or incomplete.

← Exam paper

Rarely demonstrates personal skills, perseverance or responsibility in investigative activities. Does not work within a team. Rarely approaches investigations in an ethical manner, or shows an awareness of the environmental impact. Displays very little competence in investigative techniques, generally pays no attention to safety, and requires constant supervision.

← Internal assessment

Subject reports

After every exam session, senior examiners write a subject report which is published on the IB's online curriculum centre (occ.ibo.org) and which your teacher may show to you.

The subject reports contain very useful information for you so try to get hold of a few of these. They also contain the grade boundaries for that exam. Figure 4.10 shows the grade boundaries for May 2014. Remember that this is the old ESS course; the first exams for the new course will be sat in May 2017.

Overall grade boundaries

Standard level

Grade:	1	2	3	4	5	6	7
Mark range:	0–13	14–27	28–38	39–48	49–60	61–70	71–100

Standard level internal assessment

Component grade boudaries

Grade:	1	2	3	4	5	6	7
Mark range:	0–7	8–14	15–19	20–24	25–29	30–34	35–42

Standard level paper 1

Component grade boudaries

Grade:	1	2	3	4	5	6	7
Mark range:	0–7	8–15	16–21	22–25	26–30	31–34	35–45

Standard level paper 2

Component grade boudaries

Grade:	1	2	3	4	5	6	7
Mark range:	0–6	7–13	14–19	20–26	27–34	35–41	42–65

▲ **Figure 4.10** IB ESS grade boundaries for standard level, May 2014

Each exam session will have different comments as each question is analysed but there are common recommendations nearly every time.

General advice from examiners

- Read the exam question carefully.
- Ensure you address the specific command term and actual question being asked.
- Attempt all components of the exam questions and do not leave any blank responses.
- Practise past question papers and other questions that involve application of knowledge and understanding to different situations, including mathematical calculations.
- Be familiar with the key terms and concepts in the ESS guide.
- Draw large, clear diagrams that are well labelled.
- Practise finding the links between the topics; the course is designed to be holistic.
- Use the paper in the question booklet before asking for an extra booklet. If you cannot fit your answer in the space given make it clear to the examiner where your response is continued AND, where you do continue, make it clear what question you are responding to.

- Avoid generalizations about areas of the world. These tend to be about developing countries and are insensitive, for example: Africa is a country, most babies die there, technology does not exist and so no electricity is required.
- Remember the overarching concept of international-mindedness.
- Consider global contexts .

Know where students often go wrong so you don't – check that you can do all these:
- Clarify in your own mind the differences and links between acid deposition, climate change and global warming, and ozone depletion.
- Check your understanding of environmental value systems.
- Check that you understand ecological footprints and how they differ from carrying capacity.
- Make sure you give exact detail when using examples of ecosystems.
- Show working when calculating an answer – you may then get some marks for your working even if the final answer is wrong. Always include units.
- Be concise in responding; don't be repetitive just for the sake of writing something.
- Practise drawing flow diagrams as arrows and boxes and not as pictorial diagrams – you do not have time for these.
- Practise developing a balanced argument with different viewpoints.
- Practise describing methodologies concisely, e.g. how to measure productivity.
- Practise writing multiple responses to a question, guided by the number of marks available.

Tip

Always write something

It may be obvious but you only get marks if you have written something in the response boxes. You may know an answer and be totally brilliant but if it is not on the paper, you don't get the marks. The examiner is not telepathic.

Tip

Write in the answer box

Your answer must be written in the answer box after each question. About two lines are allowed for each mark awarded so you can see how much you are expected to write.

If your writing is particularly large, you are allowed to add additional pages, but make sure that you number the extra pages carefully and clearly with the answer you are continuing. But it is far better if you only write in the answer box so think about making your writing smaller if it is very large.

iii. In the exam room

Be prepared. There is little worse than reaching the door to the exam room in a flap. You will probably be nervous but you can be in control of this. Preparation is the key.

Once you get to the day of the exam, most revision is behind you. Get to the exam room on the right day at the right time with the correct equipment and some spares.

Although you can bring any revision notes to the door of the exam room, **do not** take them inside. You do not want to be found guilty of malpractice because you forgot you had some notes in your pocket. Always check.

Once at your desk:

- Listen to the person starting the exam.
- Check that you can see the clock and have all that you need.
- If you have a question, put up your hand and wait for the invigilator to come to you. Don't stand up.
- Don't look at or try to talk to your colleagues until you leave the room at the end of the exam.
- Take some deep breaths.
- Begin.

Tip

Check your handwriting

We all now tend to type more than we write but so far, IB exams are all handwritten unless you have a special dispensation.

So can you write for hours at a time and is your writing legible?

You may need to practise your handwriting to get it fast enough and neat enough for the exam season. A good way to practise is to make handwritten revision notes.

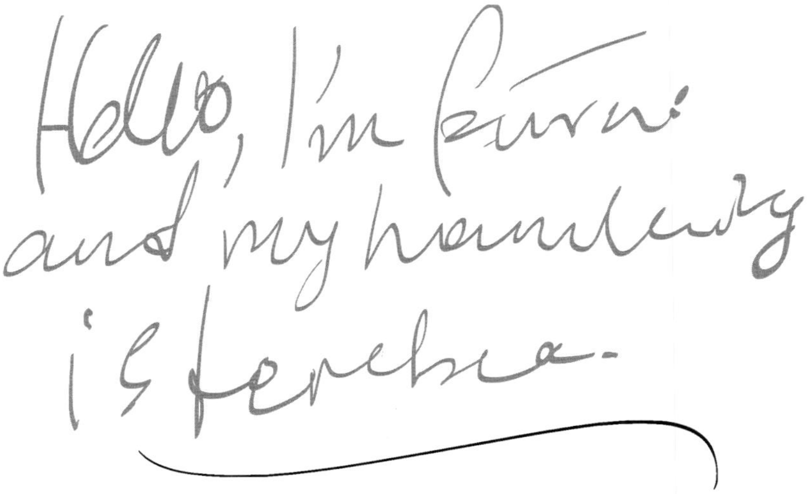

▲ **Figure 4.11** Sample of unclear handwriting – how is yours?

Approaching exams: Paper 1

Paper 1 – what is it about?

Paper 1:

- is based upon a case study in a Resource Booklet (RB) – the case study is unseen which means you do not get to see it until the exam starts
- has information that relates to a geographic location, ecosystem or ESS theme
- is worth 25% of your final marks
- has 35 marks available
- is one hour long
- has an additional five minutes of reading time before the hour starts.

What you need to do:

- answer all the questions – they are all compulsory
- demonstrate your ability to apply your knowledge from the course to novel situations
- evaluate information
- explain your viewpoints and the views of others
- justify knowledge claims
- in the final question, show you have a holistic view of the whole ESS course and can connect different topics.

How to practise Paper 1-type questions

Paper 2 of the old ESS course included a Resource Booklet.

Section A of the past papers of Paper 2 before May 2017 has questions relating to the Resource Booklet.

These will be good practice for you but remember you need to practise summarizing and evaluating the whole case study.

Case studies and local examples

Throughout the ESS guide and your course, you will be asked to use case studies. These could be local to you or global and it will really help you if you have some at your fingertips ready to answer exam questions.

List of case studies and examples to research

Here we list ideas that you could use for examples and case studies. We suggest that you work through this list either as revision practice or as part of your course. After this list are some sample case studies and examples.

Topic 1: Foundations of environmental systems and societies

1. Our environmental value systems will influence the way we see environmental issues.
   a. List other value systems that influence how we view the world.
   b. Outline one named global and one local environmental issue.
   c. Justify your viewpoint on the issues in part b and explain how your value systems influence it.

2. Look up Chief Seattle on the web. His famous speech was in the Lushootseed language, translated into Chinook Indian trade language, and then into English. While he may not have said these exact words, does it matter? Create a poster/web page on his speech and its impact.

3. Research these environmental disasters and write a short paragraph on each:
   - Deepwater Horizon oil spill
   - London smog
   - Love Canal
   - a local environmental disaster.

▲ **Figure 4.12** The Deepwater Horizon drilling rig fire in the Gulf of Mexico, 2010

4. Research these environmental movements and write a short paragraph on each:
   - Chipko movement
   - Rio Earth Summit and Rio +20
   - Earth Day
   - Green Revolution
   - Kyoto Protocol.

5. Research one local environmental pressure group or society. Create a five-year action plan for the environmental pressure group or society.

6. Look up these people who were involved in environmentalism and write three sentences on each:
   - Mahatma Mohandas Gandhi
   - Henry David Thoreau
   - Aldo Leopold
   - John Muir
   - E O Wilson.

7. Research the Millennium Ecosystem Assessment and find out the current position on the environmental indicators.

8. Calculate your own ecological footprint(EF) and compare it with the EF of others.

9. Do your own research on DDT. What evidence can you find for both sides of the argument? Be careful in looking at sources. Are they biased? Can they substantiate their claims? Do you now think that DDT should have been banned or should it still be used? Justify your opinion.

10. Find examples of human impacts on the environment and possible tipping points.

Topic 2: Ecosystems and ecology

1. Research named examples of:

   a. predator–prey relationships

   b. mutualism

   c. herbivory

   d. parasitism

   e. disease

   f. competition between species, and changes to organism abundance over time.

2. Construct models of feeding relationships, e.g. food chains, food webs and ecological pyramids, from given data.

3. Describe named examples of ecosystems you have studied.

4. Explain the transfer and transformation of energy as it flows through an ecosystem.

5. Draw examples of food chains with more than three organisms.

6. Draw examples of ecological pyramids – pyramids of numbers, biomass and productivity.

7. Research a named example of a persistent pollutant in an ecosystem, e.g. lead or mercury.

8. Describe the process of succession in a named example that you have studied first-hand. Know examples of named species from pioneer, intermediate and climax communities.

9. Describe the process of zonation in a named example that you have studied first-hand.

10. Create a key to identify eight species.

Topic 3: Biodiversity and conservation

1. Choose an animal species that is **threatened**. Browse the Red Lists or WWF if you cannot make up your mind. Find the following information:

   a. Download an image of your chosen species.

   b. What is the Red List category of your species? (It must be CR, EN or VU.)

   c. What is the global distribution of your species? Find a map if possible.

   d. What is the estimated current population size?

   e. In which CITES Appendix is your species?

   f. What are the threats facing your species (ecological, social and economic pressures on the species)?

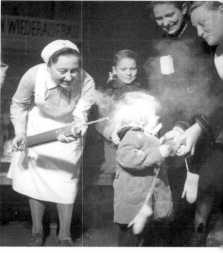

▲ **Figure 4.13** Spraying DDT on humans – it was thought to be harmless at first

g. Suggest conservation strategies to remove the threat of extinction.

h. Research how local people and government actions are involved in helping your species to recover.

Present your assignment to your class and hand out an information sheet.

2. Make sure you know the case histories of three different species:

- one that has become extinct due to human activity

- one that is critically endangered

- one whose conservation status has been improved by intervention.

For each, know the ecological, sociopolitical or economic pressures that affect the species, the species' ecological roles and the possible consequences of extinction of the species.

3. For a local and a global example of a natural area of biological significance or conservation area, describe the threats to biodiversity from human activity in these areas.

4. Evaluate the success of a named protected area.

5. Research a named example of a protected area and evaluate its effectiveness.

Topic 4: Water and aquatic food production systems and societies

1. Investigate a named local and a named global example where human activity has significantly impacted surface run-off and infiltration of water.

2. Research how two contrasting fisheries have been managed and relate to their sustainability, e.g. cod fisheries in Newfoundland and Iceland. Cover:

a. improvements to boats and fishing gear (trawler bags)

b. detection of fisheries via satellites

c. management aspects such as use of quotas, designation of Marine Protected Areas (exclusion zones), restriction on types and size of fishing gear (including mesh size of nets).

3. Research the history and current state of one of these fisheries:

- Pacific wild salmon

- North Sea herring

- Grand Banks fisheries

- Peruvian anchovy,

- or any other that is near to where you live.

4. Research with reference to a named case study, how shared freshwater resources have given rise to international conflict.

5. Research with reference to a named case study the controversial harvesting of a named species, e.g. seals, whales.

6. Research a named case study that demonstrates the impact of aquaculture and viewpoints on the harvesting of a controversial species, e.g. the Inuit people's historical tradition of whaling versus international conventions.

7. Compare a polluted and an unpolluted site (e.g. upstream and downstream of a point source) to measure aquatic pollution levels.

Topic 5: Soil systems and terrestrial food production systems and societies

1. Compare and contrast the inputs, outputs and system characteristics for two named food production systems, e.g. North American cereal farming and subsistence farming in South East Asia or intensive beef production in South America and the Masai tribal use of livestock.

2. Evaluate the relative environmental impacts of two named food production systems.

3. Research examples of the links that exist between sociocultural systems and food production systems.

4. Evaluate soil management strategies in a named commercial farming system and in a named subsistence farming system.

Topic 6: Atmospheric systems and societies

1. Evaluate the role of national and international organizations in reducing the emissions of ozone-depleting substances (ODS).

2. Research two cities with high photochemical smog levels, e.g. Hong Kong, Santiago, Mexico City or one near you and evaluate the effectiveness of management strategies.

3. Research an area where acid deposition has affected forests or lakes and evaluate the effectiveness of management strategies, e.g. the Midwest US and Eastern Canada interaction, as well as the impact of industrial Britain, Germany and Poland on Sweden.

Topic 7: Climate change and energy production

1. Make a table of these energy sources, their advantages and disadvantages and named societies that may use them:

   a. coal

   b. oil

   c. natural gas

   d. nuclear fission

   e. HEP

   f. biomass

   g. wood

   h. solar – photovoltaic

   i. concentrated solar power

   j. solar passive

   k. wind

   l. tidal

   m. wave

   n. geothermal.

2. Research an energy strategy for a named society. What energy sources are used and why? How does the society get these energy sources? What are the impacts of their use? What factors affect the energy security of the society? You could, for example, select the USA, China, Norway, Nigeria or your own country.

3. Research contrasting viewpoints on the climate change debate and justify your own viewpoint.

4. Research past and the latest international climate change talks and any agreements, and evaluate the effectiveness of the agreements.

Topic 8: Human systems and resource use

1. Compare the policies of two nation states on their approach to managing human population dynamics and growth, e.g. India, China, Singapore, Colombia, Brazil, your own country and one other. Consider cultural, historical, religious, social, political and economic factors.

2. Research an example of the dynamic nature of a resource (natural capital), e.g. whale oil use and its replacement, uranium ore, lithium or cork.

3. Select one of these examples of resource failure or choose your own. Research the reasons for it happening, the consequences and what happened afterwards.

   a. Sahel long drought 1968–72

   b. Great Chinese famine 1959–61

   c. Irish potato famine 1845

   d. Biafran famine 1967–70

4. Research examples of how exploitation of natural capital has been mismanaged by humans, e.g. hunting of some animals to extinction or near extinction – dodo, passenger pigeon, tiger, elephants.

5. Compare solid domestic waste strategies in your own town/city with those in another contrasting town/city.

6. Evaluate the ecological footprint (EF) of two contrasting individuals, e.g. yourself and one other.

Two example case studies

Case study 1: Aral Sea

See ESS guide, **4.2 Access to freshwater**.

Discuss, with reference to a case study, how shared freshwater resources have given rise to international conflict.

▲ **Figure 4.14** Map of the Aral Sea and its tributaries, notably the Syr Darya and Amu Darya (TFDD, 2007)

The Soviet Union broke up in 1991 and what had been a national issue became an international one with Kazakhstan, Kyrgyzstan, Tajikistan, Turkmenistan and Uzbekistan in the watershed of the Aral Sea and its rivers, the Amu Darya and the Syr Darya. Also small parts of the Aral Sea's basin (which was the fourth largest on Earth) or its headwaters are in China, Iran and Afghanistan.

For thousands of years, the two rivers have been used sustainably to irrigate the land. In the 1960s, the Soviet Union started transforming the region for growing cotton on a huge scale as well as melons, rice and cereals. The intensive monoculture of cotton needs high pesticide and herbicide levels. Industrial pollution added to the pollution of the area. Water quality decreased greatly. Increased irrigation of the cotton has led to the rivers' flow decreasing drastically and stopping altogether. Since 1960, the Aral Sea has lost 75% of its volume and 50% of its surface area, and salinity has tripled. However, this loss was expected once the irrigation was in place. The Aral Sea is now split into the North and South Aral Seas.

Winters are now colder and earlier and summers hotter than before 1960 as the smaller sea holds less heat. Precipitation has also fallen.

Fishing stocks have collapsed; sewage is not diluted.

Human health has suffered with higher infant mortality, more cancers and more respiratory diseases in adults.

Uzbekistan maintained the huge irrigation system set up by the Soviet Union although most water evaporated before it reached the cotton. As there was no crop rotation, chemicals were heavily used to control weeds and pests resulting in human illness. Forced labour was also used on the cotton fields.

Before the 1960 Aral Sea Plan started, some Soviet climatologists, ecologists and engineers were highly critical of the plan but environmental costs were outweighed by economic gains of growing cotton in areas of subsistence farming and low income levels.

From 1992, the nations involved and international NGOs (e.g. UNEP, World Bank and the IMF) have researched and negotiated agreements on use of the waters of the Aral Sea and its rivers but confusion and lack of funds have meant that little has improved.

Kazakhstan built a dam in 2005 to replenish the North Aral Sea. Water levels have risen by 12 m and some fishing is now viable.

The South Aral Sea is more or less drying up and is now mostly desert in Uzbekistan. Oil and gas extraction is starting there.

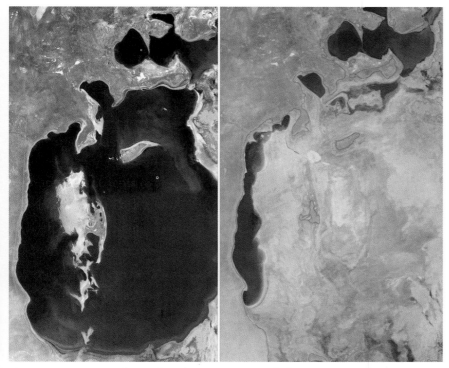

▲ **Figure 4.15** Aral Sea shrinking – 1989 on left, 2014 on right

Possible environmental solutions to the Aral Sea problem.

- Stop growing cotton.
- Grow various crops that need less water.
- Use fewer chemicals on the cotton.
- Use biological control on the cotton.
- Build dams to fill the Aral Sea.
- Build canals to channel water from the Caspian Sea or other rivers to the Aral Sea.
- Improve irrigation technology so there is less evaporation.
- Install desalination plants to decrease salinity of the water.
- Charge farmers to use the water.

Questions

1. Discuss which of these proposed solutions are reactive and which are preventative and evaluate their chance of success.

2. Describe the conflicting interests in the Aral Sea Basin. Create a table with three columns headed ecocentric, anthropocentric and technocentric. Place the possible solutions in one of the columns. Justify your choices.

3. Describe the value systems that led to the Aral Sea Plan in the 1960s.

4. Suggest reasons why developing solutions to the Aral Sea problem changed since the break-up of the Soviet Union in 1991. Has it become easier or harder to find and manage solutions since then?

5. Suggest your own solutions to the Aral Sea problem.

Case study 2: Grand Banks cod fishery collapse

See ESS guide:

1.4 Sustainability

Explain the relationship between natural capital, natural income and sustainability.

4.3 Aquatic food production systems

Discuss with reference to a case study the controversial harvesting of a named species.

8.2 Resource use in society

Outline an example of how renewable and non-renewable natural capital has been mismanaged.

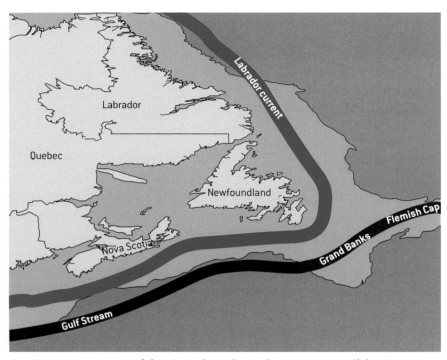

▲ **Figure 4.16** Map of the Grand Banks and its currents off the East Coast of Canada

The explorer John Cabot said in 1497 that the seas around Newfoundland were so full of fish that you only had to lower a basket into the water and raise it to find it full of fish. The reason for all the fish and crabs and lobsters was that there was a good food supply of shellfish and worms lifted from the shallow sediment by the mixing of the cold Labrador and warm Gulf Stream currents. There were haddock, swordfish and plaice, but above all, cod.

For several hundred years after that, the Grand Banks off Newfoundland were fished for cod which was salted and dried and shipped to Europe. The supply seemed sustainable with 200,000 tons of cod harvested each year and being replaced in the next year. Fishing was by gill and drift nets or longlines from small trawlers that stayed near the shore.

In the 1900s, technology changed and larger trawlers and longer lines led to a greater harvest. More profits meant more ships and fishermen and so the harvest increased to unsustainable sizes.

While it was recognized that this larger catch was unsustainable even before the Second World War in 1939, no regulation happened during the war or afterwards. By 1950 factory trawlers with freezers fished beyond the Canadian 12-mile limit, so could not be regulated, and by 1968, 800,000 tons of cod were caught. These trawlers came from Britain, Germany, Spain, France, Portugal, the USSR and even China and Japan and took the catch back to their respective countries.

It was known that this fishing caught mother cod which had been spawning for the year but nothing could be done to stop it.

By 1974, the catch was down to 300,000 tons of fish. In 1976, Canada extended its exclusive economic zone (EEZ) from 12 to 200 miles from the coast; international trawlers left but were replaced by Canadian ones whose owners wanted some of the profit now.

Dates	Catch (tons)	Type of fishing
1600–1800	8 million	Small off-shore boats
1960–1975	8 million	Freezer factory trawlers
1985	250,000	Canadian factory trawlers

▲ **Figure 4.17**

Even though fishermen warned that cod stocks were diminishing in the 1980s, fishing continued as scientific advice maintained that stocks were sustainable. Only in 1994 did a large study show that stocks were 1% of their 1960s levels and only then did the government act.

A total ban on fishing there led to large-scale unemployment of fishermen and associated workers. Some 46,000 people left the area. Processing plants reduced in size and imported Russian and Norwegian cod to process.

Commercial cod fishing had ended there.

The Canadian government was blamed for not acting sooner as were scientific advisers who may have been over-optimistic or lobbied by the fishing industry.

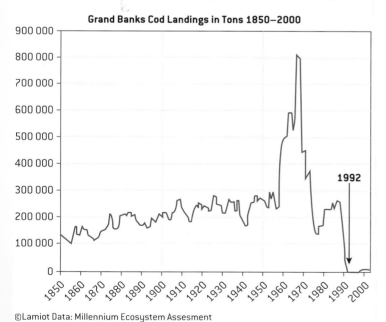

©Lamiot Data: Millennium Ecosystem Assesment

▲ **Figure 4.18** Cod catch from the Grand Banks 1850–2000

But even with the ban, cod did not return in the 1990s and early 2000s. Several reasons were proposed for this:

- Bottom trawling left the seabed so damaged and denuded that nothing can grow or live there including the crustaceans or small fish that the cod fed on.

- Trawling dispersed cod eggs so reduced birth rates.
- Capelin which were a food source for cod are not eaten any more and eat the cod eggs and larvae.
- Motherfish cod, about 10–15 years old and living further out in the ocean, had been trawled up so were not there to produce eggs.
- Seals were eating cod – which led to controversial seal culls.

By 2010, some cod were found but this was at 10% of 1960s levels and individual fish were smaller. Haddock numbers have recovered. Limited fishing continues.

Questions

1. Outline reasons why fishing the Grand Banks became unsustainable.

2. Describe one possible method of estimating fish stocks.

3. Discuss the social, economic and ecological impacts of the collapse of the Grand Banks fishery.

4. Describe the value systems probably driving this fishery in the 1960s.

5. Suggest ways that the fishery could be managed in the future to ensure its sustainability.

Paper 1 strategies and tips

Reading time

You have five minutes reading time in addition to the one hour for the exam. How are you going to use your five minutes' reading time?

In the reading time, you may not pick up a pen or write any notes but it is very useful time if you use it correctly.

First, take a few deep breaths and keep calm. There are two booklets on your desk – the Resource Booklet and the question paper. Which will you open first?

1. Opening the Resource Booklet first.

   a. Scan the pages and figures. There will be a mix of text and graphs and diagrams – about 10 figures in total.

   b. What is it about? There is a world map and then an area or theme that is the focus of the case study. Think quickly. What do you know about it? Can you visualize it in some way?

   c. Have a quick look at the diagrams and graphs. Get an impression of their focus.

   d. Scan the text and see which words jump out at you. You are absorbing far more than you imagine in the first minute of looking at the Resource Booklet and it will sit in your short-term memory while your brain makes connections about it even if you don't know it is happening.

   e. Then go to the question paper. Often the first questions are fairly easy to help you relax a bit. Remember you cannot write anything in the reading time but you can think about the answers.

   f. Then scan the questions. There are many clues in the questions. See the exam tip box to the right.

   g. If you have time left, try to match the first questions to the data in the case study.

> Browse the list of possible case studies and examples to research and decide on a local one and a global one.
>
> Research and write brief case studies on each of these.
>
> Add five questions and swap with a colleague who has written two other case studies.
>
> Discuss your answers.

> **Tip**
>
> Clues in questions:
>
> - Command term – underline or highlight it once you are allowed to start writing.
> - Number of marks allocated – think about how to get all these marks.
> - Number of lines in the box for your response – in general, there are two lines per mark.

h. Also read the final holistic question so your brain starts making connections between the pieces of data you have just seen.

2. Opening the question paper first.

   a. Scan the questions – there will be some worth one or two marks and some worth more.

   b. Remember there are 35 marks for the paper – it is fairly easy to get at least half of these.

   c. Note the command terms in the questions – what are you expected to do? You will need to make sure you do it.

   d. Then look at the case study. Can you spot some answers quickly? That should make you feel better.

   e. What is the case study about?

Think about the marks – 35 for the paper, one hour to get them. That is about 1 minute 20 seconds per mark or say about 1 minute if you allow for thinking and checking time.

You don't have to answer the questions in the order they are written. In your scanning, did you find some that you can do easily? If so, get the answers down.

Paper 1-type questions

You first see the case study when it is on your exam desk along with the Paper 1 question paper.

Past papers from between 2010 and 2016 contain examples of Resource Booklets and corresponding questions in Section A of Paper 2. That is 14 examples so plenty to use for practice. There is also a specimen Resource Booklet for this ESS course with an example of the last holistic question. Your teacher should be able to get copies of these.

Here is another example in the style of a Resource Booklet with questions to practise.

What is in the Resource Booklet?

1. There is always a world map and then enlargements of the region or area that features in the case study. Here is a sample of what one could look like.

Figure 1(a) World map showing location of Cornwall, United Kingdom

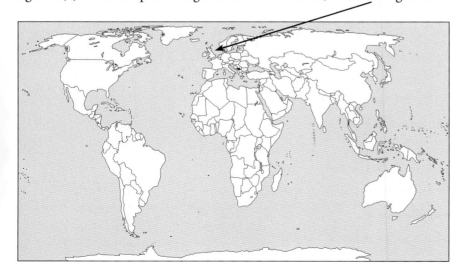

> **Tip**
>
> The Resource Booklet usually includes about 10 figures: diagrams, maps, graphs or data in fact files. It is usually not more than 1000 words so you can read it quite quickly.
>
> You are not allowed to take it out of the exam room at the end of the exam but you may write on it so have a highlighting pen with you if you want to mark information that you think is relevant to a question.

> **Tip**
>
> Remember that not all the information in the Resource Booklet will be relevant to your answers. There may be more than you need.

> **Tip**
>
> Read samples of Resource Booklets from past examinations. Although the Resource Booklet was used in Paper 2 of the old course and it is now used for Paper 1, the style and format are very similar. Try to get hold of as many as you can, as the more you see, the more comfortable you will be with them.

Figure 1(b) Map showing the United Kingdom (UK)

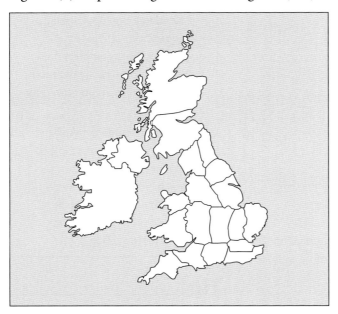

Figure 1(c) Map showing the county of Cornwall in the UK

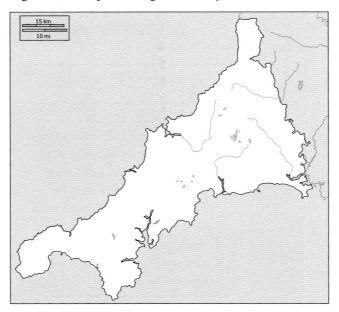

2. There will be several fact files of information. Read these carefully as they may contain some answers.

Examples of fact files

Figure 2 Fact File on Cornwall

- Rural and sparsely populated county of the UK with 144 people per km².
- One of the lowest densities of population in the UK.
- One of the poorest parts of the UK.
- Tourism contributes 25% of the economy of Cornwall.
- Agriculture and fisheries are the other main industries.
- Long coastline surrounded by the Atlantic Ocean.
- Prevailing wind from the south-west across the Atlantic Ocean.
- One of the sunniest and windiest parts of the UK.

Figure 3a Wind and solar energy in Cornwall

- Cornwall has about 400 wind turbines and 60 solar farms in operation.
- Capacity factor is the average power generated over time compared with the maximum rate.
- Solar energy is on average 12% capacity factor.
- Wind energy has a capacity factor of 30%.
- Over a year, solar and wind energy in Cornwall generates about 25% of Cornwall's electricity demand.
  Source: http://www.bobegerton.info/windandsolarinco.html

Figure 3b Maps of wind turbines and solar parks in Cornwall

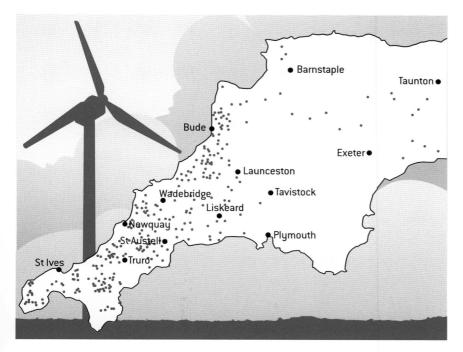

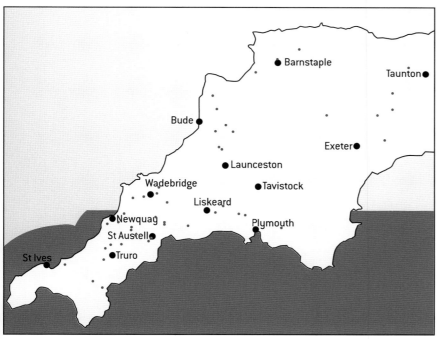

3. There is usually some graphical data. Make sure you read the graph correctly. Check the axes and key.

Figure 4 Prediction of world electricity generation sources in 2030
compared with 2011

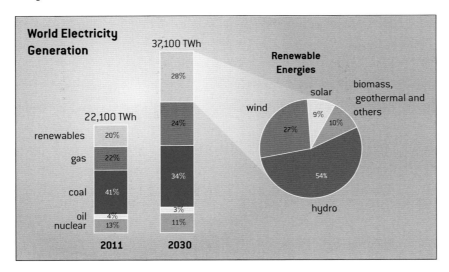

Figure 5 Wind energy in Denmark

- Wind power produced 39% of Denmark's average energy needs in 2014.

- More wind turbines per person than in any other country.

- Electricity bills are very high per household due to large subsidies for
  the wind industry.

- Before wind power, coal-fired power stations gave Denmark high per
  capita carbon emissions.

- By 2035, the target is for 84% of energy to come from wind power.

- The Danish government made planning wind farms simple and most
  involve the local community who benefit from cheaper electricity.

- When the wind does not blow, Denmark buys electricity from
  neighbouring countries.

- Due to local opposition over noise and sleep disruption, new wind
  farms are being built offshore.

Figure 6 Comparison of US and EU carbon emission targets 2000–2030

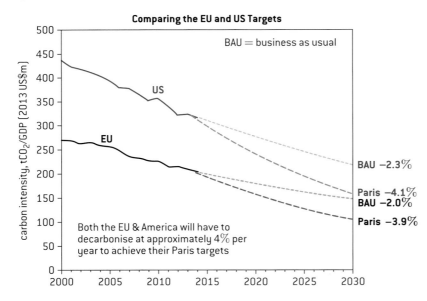

Sources: BP, International Energy Agency, Energy Information Administration, World Bank,
IMF and PwC data and analysis

Figure 7 Viewpoints on energy production

Viewpoint A

Viewpoint B

Causes of deaths of birds per year worldwide:

- Wind turbines – between 200,000 and 370,000 birds killed
- Cell phone and radio tower collisions – estimated 6.8 million
- Cats – 1.4 billion to 3.7 billion

Source:

http://www.treehugger.com/renewable-energy/north-america-wind-turbines-kill-around-300000-birds-annually-house-cats-around-3000000000.html

Tip

There are 35 marks for Paper 1. You have one hour to answer the questions. Plan how to spend the hour. Here is one idea:

Reading time	5 mins BEFORE the hour starts
Reread the Resource Booklet and questions – immediately mark the Resource Booklet sections where you think there is an answer to a question.	10 mins
Roughly 1 minute per mark, plus 5 minutes spare.	40 mins
Checking time – are you sure you answered the question? Did you leave any gaps? If you have left gaps, have a guess.	10 mins

Sample questions

Here are some questions that could be in the paper that accompanies the case study on wind energy.

1. a. Identify two sources of renewable energy other than wind. [1]

    b. Outline two advantages of wind energy. [2]

2. a. With reference to Figure 5 and your own knowledge, explain two reasons why wind energy may be less effective than energy from oil. [2]

    b. Evaluate the use of wind energy in Denmark. [4]

3. With reference to Figure 6 and other information in the Resource Booklet:

    a. Compare the decarbonizing rates of the US (America) and EU (European Union). [2]

    b. Suggest two reasons why BAU predictions are higher than the Paris targets. [2]

4. Our opinion on the use of renewable energy sources depends on our environmental value system. State in the table below examples of a technocentric and an anthropocentric response to use of wind turbines. [2]

Technocentric	Anthropocentric

5. With reference to the information in the Resource Booklet and your own knowledge, evaluate the significance of renewable energy sources in the world energy supply. [6]

These sample questions and sample case study figures are not official IB ESS documents but do give you a taste of the style to expect.

Tip

Question 1 has text boxes to show you how your exam paper will look. You write your answers inside the box. If you need more space, do not go outside the box. Instead continue your answer on another page and mark the question number very clearly. Make sure you indicate that you have continued your answer elsewhere – the examiner needs to know there is more.

Tip

Remember the final question in Paper 1 will be worth about 6 marks out of the 35 and you will be expected to take a holistic view in your answer. See question 5 on this page. That means pulling several concepts together and planning out your answer before you start writing, just like a mini-essay.

Paper 1 questions and answers

Get a copy of the May 2010 exam papers for ESS.

The case study is in the Resource Booklet of Paper 2. It is about the island of Borneo.

Read the case study and the Paper 2 questions and markscheme.

Now let's go over some of the questions.

1. (a) (i) The island of Borneo lies on the equator. State **two** types of ecosystem that may be found on Borneo. *[1]*

   ....

   ....

   (ii) Explain why gross primary productivity is high in Borneo. *[1]*

   ....

   ....

   ....

   ....

   (iii) Outline **two** reasons for the high levels of species diversity in Borneo. *[1]*

   ....

   ....

▲ **Figure 4.19** May 2010 Paper 2, question 1a

You may be able to answer these three questions without needing the case study as the question tells you it is on the equator.

Note also the command terms – state, explain, outline.

And the marks – 1, 1, 1.

You know from topic 2.4, biomes, zonation and succession, that:

- The tricellular model of atmospheric circulation explains the distribution of precipitation and temperature and how they influence structure and relative productivity of different terrestrial biomes.

And in the guidance section of 2.4:

- Students should be encouraged to study at least four contrasting pairs of biomes of interest to them, e.g. temperate forests and tropical seasonal forest, or tundra and deserts, or tropical coral reefs and hydrothermal vents, or temperate bog and tropical mangrove forest.

So the answer to a(i):

- could be tropical rainforest/coral reefs/mangrove swamps/alpine forest/montane forest/tropical grassland/savanna/cloud forest/palm oil plantation
- but not terrestrial/aquatic/marine/forest/river/ocean/riverbank/coast, as these terms are too vague.

(A forward slash (/) here means these are alternatives.)

For a(ii), you also know from 2.4 that:

- Each of these classes will have characteristic limiting factors, productivity and biodiversity.
- Insolation, precipitation and temperature are the main factors governing the distribution of biomes.

So the answer to a(ii) is:

- high because solar radiation/temperature/light and water are abundant/water and light not limiting
- or something like: favourable climatic conditions lead to increased photosynthesis/primary productivity.

To answer a(iii), you need to think about the ecosystems on Borneo and productivity. Any of these answers will gain you the mark:

- high level of producer biomass/more food
- island where speciation has occurred due to natural selection/isolation from mainland species
- many ecological niches/food webs in rainforest
- mature ecosystems so time to develop complexity/stability
- variety of ecosystems due to differences in altitude/habitat
- lack of human activity (until recently) has meant that diversity has built up over time.

Where might you miss marks?

Part (i) could be a two-word answer as all you do is to **state**.

Did you **explain** in part (ii)? You need a reason here and note that there are four lines instead of two for the mark which suggests you have to write more.

Part (iii) also needs reasons even though the command term is **outline**. That means that writing that it is a mature ecosystem or has a high biomass is not enough to gain the mark as you need to give reasons as well.

Now look at the last question, question 1(f):

1. (f) (i) Evaluate the message in the slogan " Save Borneo Save The World" on the poster in Figure 10. [3]

....

....

....

....

....

....

(ii) Justify your personal viewpoint on the value of international cooperation in the conservation of tropical rainforests. [2]

....

....

....

....

▲ **Figure 4.20** May 2010 Paper 2, question 1f

This has command terms 'evaluate' and 'justify' which are both objective 3 terms – so they are meant to be more challenging.

In f(i), to evaluate you need to appraise the slogan by weighing up its strengths and limitations. In this case, you might write something like:

> One limitation of the poster is that saving Borneo will not save the world/ what does save the world mean?/is it save the human species? [1]

> But a strength is that Borneo is an example of what we can do/if we can save Borneo from loss of species diversity/loss of rainforest, we can do the same elsewhere. [1]

> And as Borneo is the third largest island in the world, such a large area could make a difference to climate change if reforested. [1]

> Borneo provides ecosystem services/which benefits the world, e.g. high biodiversity/important part of water cycle/large carbon store. [1]

> The tree with the leaves holding the world identifies the importance of the rainforest. [1]

> It looks like the world is one big island/all interconnected/ interdependent, which is a very good message. [1]

> But people might find it hard to relate to the message/slogan without an explanation. [1]

There are more than 3 marks here as these are alternative points. As long as you have a strength, a limitation and one other point, you get the full [3] marks.

In f(ii), you not only have to put your viewpoint but also to say why it is held. Something like:

- Markets for rainforest products are global; only by limiting these can the resources be managed sustainably. [1]

Or:

- Transnational corporations (TNCs) are international organizations and have the ability to coordinate global markets.[1]

And for the other mark, something like:

- MEDCs have the funds to respond to conservation needs in LEDCs so could support them/MEDCs have often exploited resources from LEDCs so should help conserve what is left. [1]

Or:

- Small scale/local projects can be more effective at bringing about change, e.g. Samboja. [1]

There are many other acceptable and reasonable suggestions. If you argue the other way, that international cooperation does not work, that is acceptable too.

But you would not get the marks if you argue for/against the conservation of rainforests as that is not an answer to the question.

Now try to answer all the other questions in this May 2010 paper and compare your answers with the markscheme.

Then try some other past papers and case studies.

Tip

- Check the mark allocation, space available and command term to help you answer correctly.

- There are many ways of saying the same thing which is why examiners need to think about what you are saying and decide if you get the mark or not – they are humans, not machines.

- Think before you write so that your answer is as clear as possible.

Approaching exams: Paper 2

Paper 2 – what is it about?

Remember that Paper 2 represents 50% of your marks for ESS, so you need to get this one right. It is 2 hours long with 65 marks available in total.

Paper 2 has two sections.

Section A is compulsory short answers and data-based questions. It is marked out of 25.

Section B gives you four essays from which you must choose two. This is the only choice you get in ESS exams. Each essay is worth 20 marks. There are three sections, with marks usually allocated as:

a. 4 marks

b. 7 marks

c. 9 marks.

All questions are marked in accordance with the markscheme except the final part of each essay in Section B, which is marked using markbands. Markbands are a statement of the performance expected and your answers are judged against this. A best-fit is used to find the descriptor that most accurately fits your answer. See figure 4.21 for the markband descriptors.

Marks	Level descriptor
0	The response does not reach a standard described by the descriptors below and is not relevant to the question.
1–3	The response contains • minimal evidence of knowledge and understanding of ESS issues or concepts • fragmented knowledge statements poorly linked to the context of the question • some appropriate use of ESS terminology • no examples where required, or examples with insufficient explanation/relevance • superficial analysis that amounts to no more than a list of facts/ideas • judgments/conclusions that are vague or not supported by evidence/argument.
4–6	The response contains • some evidence of sound knowledge and understanding of ESS issues and concepts • knowledge statements effectively linked to the context of the question • largely appropriate use of ESS terminology • some use of relevant examples where required, but with limited explanation • clear analysis that shows a degree of balance • some clear judgments/conclusions, supported by limited evidence/argument.

7–9	The response contains ● substantial evidence of sound knowledge and understanding of ESS issues and concepts ● a wide breadth of knowledge statements effectively linked with each other, and to the context of the question ● consistently appropriate and precise use of ESS terminology ● effective use of pertinent, well-explained examples, where required, showing some originality ● thorough, well-balanced, insightful analysis ● explicit judgments/conclusions that are well-supported by evidence/argument and include some critical reflection.

▲ **Figure 4.21** Paper 2 Section B markband descriptors (from IB ESS guide, February 2015, page 76)

The idea is that the examiner marks your part (c) and then reads the markband descriptors to find the one that most accurately describes how you have done. Which mark you get in a markband depends on the extent to which your work shows the qualities listed.

Paper 2 strategies and tips

How will you approach Paper 2? It is two hours long with five minutes' reading time before the two hours starts.

Tip

Watch the clock

As in all exams, if you don't write it down, you can't get any marks for it. There is no point walking out at the end of the exam thinking that you knew something but ran out of time to get it down on paper.

So, plan, plan, plan your timing and stick to it.

Section A is worth 25 marks; each essay of Section B is worth 20 marks. You need time to decide which essays to do, plan your answers and check your paper at the end.

One suggestion for how to allocate your time is:

Total time	120 minutes	
Planning	10	Decide which essays to do – choose two out of the four on the paper, plan them
Section A	35	Scan and do the 'easy' questions first
Essay 1	35	Do your second best first?
Essay 2	35	Don't run out of time
Checking	5	Read through – don't bother about spelling, check if you can add a few marks

Tip

What will you answer first?

Remember, you do not have to answer the questions in the order they appear on the paper. You could even do Section B, the essays, first but do you want to?

In Section A, you could answer the easier questions first. Check the command terms. If they are objective 1 terms – define, draw, list, label, measure, state – or some objective 2 ones – outline, identify – they may be easier marks to gain.

Getting a few marks and having some answers done should give you more confidence to tackle the rest.

Section A – short response questions and answers

This section (25 marks) tests your knowledge and understanding of the whole course and tests if you can make connections between various topics and apply concepts to new data.

Try to answer this question (May 2011 Paper 2, question 1 (a–f)).

Figure 1 below shows age/sex pyramids (X, Y and Z) for three different countries in the year 2010.

Figure 1

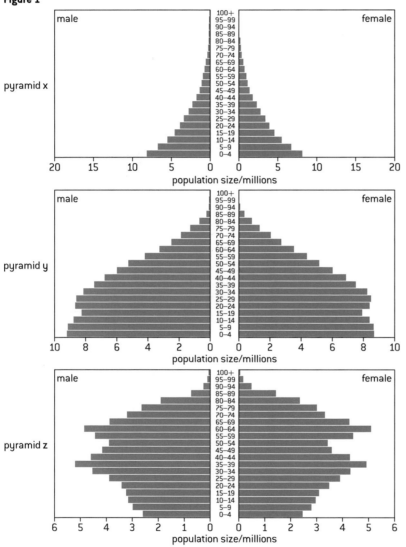

Source: www.census.gov/ipc/www/idb/pyramids.html

(a) State which pyramid (X, Y or Z) represents each of the following countries. [1]

Brazil:	....
Ethiopia:	....
Japan:	....

Figure 2 below shows the demographic transition model.

Figure 2

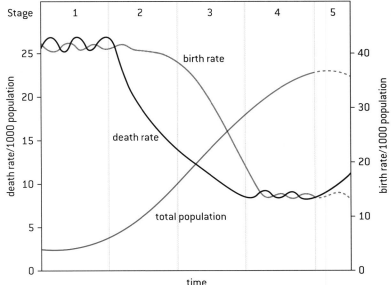

Source: http://i.ehow.com/images/GlobalPhoto/Articles/2243559/Demographpic TransitionModel-mail Full.jpg

(b) Identify the stage of demographic transition in which you would expect to find
each pyramid in Figure 1. [1]

Pyramid X: ....

Pyramid Y: ....

Pyramid Z: ....

(c) Define the term *ecological footprint*. [1]

(d) State how the ecological footprints of a country in stage 1 and a country in stage 4
of the demographic transition mode (Figure 2) would differ. [1]

(e) Explain **three** reasons for the difference in ecological footprints you have identified
in part (d). [3]

(f) Evaluate the concept of an ecological footprint as a way of measuring resource use.   [2]

> ....
> ....
> ....
> ....

How did you do?

Here are some answers:

(a) State which pyramid (X, Y or Z) represents each of the following countries.   [1]

Brazil:	Pyramid Y
Ethiopia:	Pyramid X
Japan:	Pyramid Z

For parts (a) and (b) you need to know 8.1, human population dynamics.

- Age/sex pyramids and demographic transition models (DTM) can be useful in the prediction of human population growth. The DTM is a model which shows how a population transitions from a pre-industrial stage with high CBR and CDR to an economically advanced stage with low or declining CBR and low CDR.

> If you don't know the answer – guess!

(b) Identify the stage of demographic transition in which you would expect to find each pyramid in Figure 1.   [1]

Pyramid X:	stage 1/stage 2/stage 1 or 2
Pyramid Y:	stage 3/stage 4/stage 3 or 4
Pyramid Z:	stage 4/stage 5/stage 4 or 5

> Only 1 mark available for this so you need to get them all correct. No half marks will be given.

(c) Define the term *ecological footprint*.   [1]

> The area of land (and water) required to support a human population/individual at a given standard of living/indefinitely/sustainably and absorb all their waste.

> Part (c) is recall of a definition of a term.

(d) State how the ecological footprints of a country in stage 1 and a country in stage 4 of the demographic transition mode (Figure 2) would differ.   [1]

> Stage 1 EF will be smaller than stage 4 EF

> Part (d) requires you to state an answer so this is enough for the mark.

Parts (c) and (d) are covered in topic 8.4:

- EF is the area of land and water required to sustainably support a defined human population at a given standard of living. The measure takes into account the area required to provide all the resources needed by the population, and the assimilation of all wastes.

- EF is a model used to estimate the demands that human populations place on the environment.

- EFs may vary significantly from country to country and person to person and include aspects such as lifestyle choices (EVS), productivity of food production systems, land use and industry. If the EF of a human population is greater than the land area available to it, this indicates that the population is unsustainable and exceeds the carrying capacity of that area.

- **Compare and contrast** the differences in the ecological footprint of two countries.

(e) Explain **three** reasons for the difference in ecological footprints you have identified in part (d). [3]

> Different diets as people in stage 4 countries tend to eat more meat/people in stage 1 tend to eat more vegetables and cereals.
> Stage 4 higher as uses more energy as more appliances/cars/finance for these.
> Stage 4 higher as uses more transport/travel and so more carbon emissions.
> Stage 4 higher as more imported goods so more poilutants/food miles.

> Part (e) requires reasons in 'explaining'. There are 4 marks available for this answer so it is a good idea, if you have time, to write more in case one of your reasons is not accepted.

(f) Evaluate the concept of an ecological footprint as a way of measuring resource use. [2]

> Strengths – useful snapshot of sustainability of a population.
> Or a tool for governments to measure sustainability and persuade population to change.
> Limitations – does not include all information as only a model.
> Estimate so not accurate.
> Does not show types of resources.

> Part (f) asks you to 'evaluate' – remember both strengths or advantages and limitations or weaknesses.

Parts (e) and (f) also require knowledge of topic 8.4 as well as of 1.2, systems and models:

- A model is a simplified version of reality and can be used to understand how a system works and predict how it will respond to change.

- A model inevitably involves some approximation and loss of accuracy.

- **Evaluate** the use of models as a tool in a given situation, e.g. for climate change predictions.

> 1. Find past exam papers and try the Paper 1 questions which will be similar to the Paper 2 Section A ones that you will have to answer.
> 2. Start slowly.
>   For the first paper you try, have your textbook and notes available. If you can't answer a question, look it up.
>   Do not time yourself for this paper.
>   Then check the markscheme and mark your answers.
> 3. For the second paper you try, close your books and have a go but do not look at the clock. Then check and mark as before. Look things up now if you got it wrong.
> 4. For the third and subsequent papers, time yourself strictly to one hour. Check and mark.

Section B – essay-style questions and answers

Remember, you select two essays out of the four given on the paper. Don't do only one or more than two. Spend about 35–40 minutes on each one. Getting half marks (10/20) is relatively simple as long as you answer the questions. It is getting the other 10 that can be a challenge.

Here is an older Section B question (so the mark allocation is not 4, 7 and 9) but it will give you an idea of what a good answer looks like.

May 2009 Paper 2, question 5

5. a. Outline the link between greenhouse gases and global temperatures. *[2]*

    b. Explain why the effects of global environmental problems, such as global warming and ozone depletion, will not have an impact on every society to the same extent. *[6]*

    c. Human responses to global warming can be divided into strategies to prevent global warming from happening (preventive) and strategies to reduce the impacts that global warming might have (reactive). Outline preventive and reactive management strategies to address global warming and evaluate their strengths and weaknesses. *[10]*

(This has been replaced by 20 marks available for each essay: (a) 4; (b) 7; (c) 9; and using the markbands to assess part (c) of the essays.)

And now the same question but with examples of good answers to each part and the associated marks.

5. a. Outline the link between greenhouse gases and global temperatures. *[2]*

> (a) Greenhouse gases help to insulate the Earth by retaining longwave radiation and maintaining temperatures. They prevent energy from leaving the Earth. [1]
>
> Increases in the concentration of greenhouse gases in the atmosphere are believed to cause increasing global temperatures. [1]

> This is in topic 6.1, atmospheric systems and societies:
> * The greenhouse effect of the atmosphere is a natural and necessary phenomenon maintaining suitable temperatures for living systems.
> * Human activities impact the atmospheric composition through altering inputs and outputs of the system. Changes in the concentrations of atmospheric gases such as ozone, carbon dioxide, and water vapour have significant effects on ecosystems.
> * **Outline** the role of the greenhouse effect in regulating temperature on Earth.

    b. Explain why the effects of global environmental problems, such as climate change and ozone depletion, will not have an impact on every society to the same extent. *[6]*

Tip

Planning the essays

You write the exam in an answer booklet and can ask for extra pages to add to this if you fill it up. Think about planning your answers to the essay questions IN the booklet, perhaps as a mind map. It does not matter how rough this plan is as you will write the essay and then draw a line through the plan. But if you should run out of time, you can hand in the plan and at least get a few marks for this.

Straightforward knowledge and understanding need to be demonstrated and only 2 marks are available so write concisely and move on.

(b) Some societies are more affected than others due to their geographical location. [1]

Increased sea levels will not be an issue for landlocked countries but could cover and wipe out low-lying island nations, e.g. the Maldives, Tuvalu and low-lying areas of Bangladesh. [1]

LEDC countries, e.g. Somalia, may have decreased rainfall and so fall in agricultural output and famines and droughts. MEDCs with reduced rainfall may be able to irrigate more or import more food. [1]

Countries that are heavily glaciated areas may suffer more from melting/flooding with increase in global temperature, e.g. Scandinavian countries. [1]

Greenhouse gases and ODS are released from some countries more than others but are carried in the atmosphere all around the Earth.

The ozone layer is thinner in higher latitudes where MEDCs with better health and education may be able to protect population better against skin cancers from increased UV. [1]

MEDCs with greater wealth and technological development can mitigate against climate change more readily (through constructing sea defences/hurricane warning systems). [1]

Some countries may benefit if temperatures rise, e.g. Canada and Russia could mine for oil under the Arctic Ocean if it melts.

> There are more points here than marks available so this is a really good answer but the student has written more than needed to get the maximum marks. This is acceptable if you can write quickly and legibly. If you write slowly try and stick to the right number of points to get the marks.

c. Human responses to global warming can be divided into strategies to prevent global warming from happening (preventive) and strategies to reduce the impacts that global warming might have (reactive). Outline preventive and reactive management strategies to address global warming and evaluate their strengths and weaknesses. *[10]*

> This is the most challenging part of the question and has the most marks so only answer this question if you think you can do part (c).

(c) **Preventive management strategies** stop climate change by tackling the reason and causes. These include:

reducing emissions of carbon dioxide, methane, HCFCs/HFCs, nitrogen oxides through changes to current practice, [1]

reducing fossil fuel use, e.g. through developing alternative energy sources, e.g. renewables/nuclear or using technology to improve efficiency, [1]

energy efficiency, e.g. in buildings/transport/changing farming practice, e.g. reducing cattle farming (methane)/artificial fertilizer use, [1]

incentives provided through international targets/carbon taxes,

increasing natural recycling of carbon, by reducing deforestation/increasing afforestation,

high technology solutions, e.g. climate engineering/phytoplankton farms/carbon sequestration,

plans to suspend small mirrors in space between the sun and the Earth to deflect solar radiation,

plans to add sulphur to jet fuel to increase sulphur dioxide in atmosphere (adding particulates) so reducing solar radiation reaching Earth's surface, attempts to limit the human population.

> There are two parts to this – outlining the strategies and then evaluating them. You will have to guess how the marks are split between them. It could be 5 + 5 so make sure you do both.

> There are more points made than marks available here, but be careful if you know a lot – only put it down if you have time. Otherwise, check your timing and come back to this at the end if you want to say more.

Reactive strategies deal with the impact and effects of climate change: e.g. engineering works to protect coastal areas from flooding, [1]

improving prediction and warning systems to reduce impact of increased natural hazards, e.g. hurricanes, [1]

migration of people to cooler/wetter areas, [1]

land-use planning, reducing the densities of people living in most vulnerable areas,

contingency planning, investing in emergency services/stockpiling food stores to offset disaster in emergency situations.

In evaluating this, preventive might be more important because this is trying to stop the problem happening in the first place. [1]

However, reactive might be more important, because even if we stop releasing greenhouse gases today, the argument is that the effects will still be felt from gases emitted in the past and all we can do now is to mitigate the impacts. [1]

Reactive might be more cost effective, because we are not sure exactly what effects will be felt where, and so money can be targeted when real problems emerge instead of spending it if we don't need to. [1]

Reactive means we don't have to change our behaviour now but wait and see what happens which may make things worse in future.

Preventive may be more important because if we reduce greenhouse gas emission now we can offset the worst effects of climate change in the future.

Preventive might be more important because they depend on international cooperation, and countries working together are more likely to bring about effective change than small changes in single countries, e.g. Denmark with wind energy, recycling in Germany. [1]

Here is another essay-style question.

May 2011 Paper 2, question 3

3. **a.** Define the term *biodiversity*, and explain how species diversity for an area may be calculated. *[4]*

   b. Identify the ways in which unsustainable agricultural practices may lead directly **and** indirectly to a loss of biodiversity. *[5]*

   c. Evaluate the relative importance of factors that determine the sustainable use of freshwater resources. Refer to at least **one** case study in your answer. *[9]*

Expression of ideas: [2 max] Total: [20]

> You may see this 'Expression of ideas' mark in older past exam papers. The 2 or 3 marks were designed to reward candidates who wrote good essay-style answers, not lists, with examples and clear discussion of points. Now replaced with the markband criteria.

3. (a) *Biodiversity* is biological/living diversity per unit area. It is made up of species, habitat and genetic diversity. [1]

Take a transect/quadrat/sampling method to collect data on number and abundance of species. [1]

Simpson's/a diversity index is used to calculate (species) diversity. [1]

Then values are extrapolated for larger areas. [1]

> But the Lincoln index would be a wrong answer as it is for calculating a population of motile organisms, not biodiversity.

(b) Agricultural practices are unsustainable if they damage the natural capital/reduce the ability of the environment to yield crops/produce in the long term. [1]

or

It is farming in a way which damages the environment.

Ways in which there is a direct loss of biodiversity include: monoculture can lead to loss of species diversity, [1]

pesticides can kill all species not just the species being targeted, [1]

making silage/harvesting grass before flowering reduces pollination/seed production,

deforestation leads to loss of habitat diversity.

Ways in which there is an indirect loss of biodiversity: genetically modified (GM) crops can reduce species diversity as natural grasses are outcompeted, [1]

removal of hedgerows to increase farm sizes can lead to habitat loss, [1]

draining wetlands to provide more farmland can lead to the loss of habitats,

excessive application of inorganic fertilizers can lead to eutrophication of nearby water bodies and loss of species,

deforestation leads to loss of habitats and food sources for other organisms causing them to become extinct,

native species are outcompeted by domestic species (which have escaped captivity).

> There are 5 marks available for part (b) and far more points are given in this answer. Worth doing if you know more and have time but not if you spend too long on it.

> The categories direct and indirect need to be clearly stated to get the marks.

c. Evaluate the relative importance of factors that determine the sustainable use of freshwater resources. Refer to at least **one** case study in your answer. *[9]*

Before you read the answer to (c), draw up a rough plan of how you would answer it. It is worth 9 marks so how are you going to get these? Remember what evaluate means – consider the strengths and weaknesses or advantages and disadvantages. What case study do you know? What does sustainable use mean? Quickly write down what you recall about it. List the factors and number them roughly in order of importance. How many factors do you want to talk about? Remember the whole question should take no more than 40 minutes.

(c) *Sustainable use* means using water at a rate which allows natural replenishment/regeneration, or using water in a way that minimizes damage to the environment. [1]

Factors that determine this include inequity of use and: political conflict, population size and growth, migration levels, levels of industrialization, increased food production/irrigation, socio-economic levels and development levels of the population, cultural attitudes towards conservation/environmental education, aquifer sizes/depths and distribution, climate/precipitation/aridity/drought, global warming/climate change, available technology. [max 3 marks]

The relative importance of different factors will depend on the context of the are a, e.g. in semi-arid/desert areas there is a smaller stock of water resources in the first place, conflicts over water can be made worse when there are political tensions between different user groups, or in societies with a short-term attitude toward resource use. In countries near to carrying capacity, population numbers will be more significant than in countries where water supplies are abundant. In countries where water is abundant it may be perceived as an unlimited resource and therefore wasted. In societies where there is good awareness of sustainability issues, individuals may take steps to ensure their own level of water use is sustainable, e.g. recycling rain water. In countries with oil resources technological solutions to water shortages such as desalinization are a more affordable option but in LEDCs expensive high tech solutions are less of an option.　　[max 3 marks]

e.g. Aral Sea – short-term gain in cotton industry at expense of sustainable water use or

e.g. Colorado river – drastically reduced flow due to high consumption/ wasteful abstraction of groundwater resources.　　[max 2 marks]

So you see many possible answers, there are more than needed for full marks here.

Note how the marks are divided up – 1:3:3:2. Make sure your answers cover all parts of the question.

Expression of ideas: [2 max] Total: [20]

Now reread the markband descriptors for part (c). How do you score this one? Remember, the final mark out of 9 comes from the markscheme and then the markband descriptors. Is it in the top band 7–9? Yes. Would you give it 7, 8 or 9? Why?

Part (c) will probably be the most challenging part of Section B so it is worth practising it some more.

State whether you believe global warming **or** biodiversity loss represents the bigger threat in the future. Justify your viewpoint.　　[9]

May 2012 Section B, question 4c

(c) Answer 1 – arguing for global warming as greater threat

Global warming is a bigger threat than biodiversity　　[1]

because it will have big effects, e.g. melting ice caps/shifting biomes/ changing the world's weather/flooding land. [1] Its effects will be more widespread than biodiversity as every country will be affected [1] whereas the loss of diversity in a tropical rainforest, for example, may have little direct impact on people elsewhere.　　[1]

Larger numbers of people will be affected, e.g. Bangladeshi delta region low-lying and densely populated affected by sea level rises. [1] It will affect a bigger range of human activity, such as food supply/living space, health/disease. [1] It will lead to biodiversity loss as well (whereas loss of diversity will not cause global warming). [1] Many biodiversity hotspots are particularly vulnerable to climate change, e.g. coral reefs. [1] It has a higher public/global profile and it can lead to significant social consequences, e.g. wars/mass migrations. [1] It is harder to solve than biodiversity loss which can be helped by, for example, seedbanks. [1] The current rate of warming is unprecedented, whereas there have been mass extinctions in the past and the biodiversity has recovered. Global warming may lead to positive feedback loops/potentially devastating tipping points/exponential change.

(c) Answer 2 – arguing for biodiversity loss as greater threat

Biodiversity is a bigger threat than global warming [1]

because biodiversity provides/ensures key essential ecological services, [1] e.g. a balance of atmospheric O_2 and CO_2/soil preservation/mineral cycles. Without these there would be no life-supporting conditions on the planet. [1]

Once biodiversity is lost it is gone forever. [1] Whereas global warming may be mitigated/reduced by negative feedback mechanisms, there are no feedback mechanisms to maintain diversity. [1]

To restore biodiversity it can take very long periods of time whereas climatic variations are reversible/happen in cycles (and have occurred in the past). [1]

From a human point of view, loss of biodiversity may mean loss of critical resources.

Biodiversity provides many forms of natural income, e.g. aesthetic, economic, ecological services at a point in human population growth where resources are possibly limiting. [1]

It is biodiversity that gives living systems the ability to adapt to change and the more it is reduced, the lower the chances of adapting to change. [1]

Biodiversity is being lost now/threats seem more immediate to people.

There are some positive benefits of global warming but no benefits to biodiversity loss. [1]

> Plan your answer first. Will you argue for global warming or biodiversity being the greater loss? You can argue for either.

> Define what you mean by bigger. Is it the scale of the threat, the likelihood of it or if it threatens humans or all living systems?

> You need to include comparisons with the other threat. If you select one and just describe it, you will not get more than about half of the available marks.

Read the question and answer and decide which markband it falls into. Below is another example of this type of question and some hints on how to approach it.

(c) Explain why some people believe that the ecological footprints of some countries need to be reduced. Justify whether an ecocentric **or** a technocentric approach to reducing the ecological footprint is more likely to be successful. [9]

May 2012 Section B, question 3c

1. Plan your answer. Think about including these:
   - Definition of ecological footprint.
   - Why some think it needs reducing.
   - Examples of countries with large and smaller EFs – it always helps to think about a real example.
   - Decide whether you support ecocentric or technocentric – you need to argue for one and put the argument for the other as well.
   - Think of examples to illustrate your points.
2. Write your answer in 15 minutes.
3. Mark your answer using the markscheme on the companion website.
4. Check the markbands.
5. Now practise other part (c) questions in the same way.

Extended essay – the basics

From September 2016, there will be a new guide containing new assessment guidance for the extended essays in all subjects. Make sure you have a copy of this guide as it contains information that will help you to write a good EE. Here are some key facts about the EE:

- It is compulsory for all Diploma students.
- You must achieve a D or above to be awarded the Diploma.
- Course students can write an EE if they choose.
- It develops some very useful skills for university study.
- It is externally assessed which means it will be marked by an IB examiner.
- Along with your TOK grade it can gain you up to 3 additional marks towards your total Diploma score.
- It is a 4000-word piece of independent research in a Diploma subject.
- It leads to a formally presented piece of structured writing.
- It is presented as a formal piece of academic writing of 4000 words (maximum) plus a reflection of up to 500 words.
- It is a piece of independent research on a topic of your choice.
- It is under the supervision of a supervisor who is a teacher in your school.
- You also check in with your supervisor to give informal updates.
- It should take approximately 40 hours of work.
- Supervisors should spend around 3–5 hours supporting you and that includes the three reflection sessions.
- There is a mandatory reflective viva voce with your supervisor at the end of the process.
- It is assessed against criteria that are common to all subjects.
- We recommend that you do your EE in a subject you are taking as that gives you sufficient subject knowledge.

The EE is a reflective and independent piece of work. You will undertake three formal assessment sessions (including a viva voce at the end of the process).

What the EE can do for you

As you research and write your essay, you will develop a number of skills that will be of great use to you for your further studies. You will:

- conduct an in-depth investigation on a research topic of particular interest to you
- will receive the guidance of a supervisor
- improve your academic research, intellectual discovery, and creativity
- improve your writing and communication so that you present the findings of your research in a reasoned and coherent manner that is appropriate to the ESS
- develop your reflective skills.

Being reflective is one attribute of the IB learner profile. Reflection in the EE focuses on your progress during the planning, research and writing process. The emphasis in the EE is on the process of reflection. That means you reflect on conceptual understandings, decision-making, engagement with data, the research process, time management, methodology, successes and challenges, and the appropriateness of sources. You are encouraged to informally reflect throughout the experience of research and writing the EE, and to reflect formally during the reflection sessions with your supervisor and when completing the Reflection on Planning and Progress form (RPPF).

What to expect

Inquiry and the extended essay

Your extended essay is perhaps the most inquiry-based activity that you undertake in the Diploma. It is an open-ended task and perhaps the first where you direct the topic. But you are not alone. You have your supervisor, school librarian, DP coordinator and teachers to help you.

What your school does

Your IBDP or EE coordinator will ensure that:

- Your essay conforms to the regulations.
- You have a supervisor that can help you in ESS.
- You have the following documents/forms:
  - the general and ESS subject-specific information and guidelines that are necessary
  - Reflection on Planning and Progress form (RPPF) at the start of the process and that it is completed and signed at the end
  - EE criteria
  - the most recent ESS EE subject report.
- You have exemplars to see what makes a good (and bad) EE in ESS.
- You are familiar with the academic honesty policy and how to cite and reference.

What to expect from your supervisor

The relationship you have with your EE supervisor will be very important as they play a major role in helping you undertake your research for the EE. The relationship is a two-way process: the supervisor provides support but you must come with ideas and take an active role in the reflection sessions. You should initiate discussions and arrange meetings.

Your supervisor is responsible for:

- discussing the choice of topic and helping you formulate a well-focused research question
- ensuring that the research question satisfies legal and ethical standards, health and safety, confidentiality, human rights, animal welfare and environmental issues
- explaining the role of the supervisor so you understand what to expect from them

> **Tip**
>
> There are certain documents that you should be aware of and read before undertaking an EE in ESS. Your supervisor or Diploma coordinator can provide you with these IB documents that include:
>
> - EE guide
> - Academic honesty guide
> - IB Ethical guidelines
> - IB Animal experimentation policy
> - Effective Citing and Referencing
> - Exemplar EEs in ESS
> - EE subject reports from past exam sessions.

- ensuring you have all the relevant documentation you need (see 'What your school does' above)
- discussing with you the:
  - nature of the EE
  - topic and research questions
  - appropriate research methods
  - appropriate resources – library, people, laboratory etc
  - method to use to cite and reference
  - fact that Diploma candidates must score a D or above to qualify for the Diploma.
- conducting three reflection sessions with you. After each session the RPPF must be completed, signed and dated by them and you
- monitoring your progress and offering guidance to make sure the work is your own
- reading and commenting on **ONE** complete draft – they must NOT edit the draft
- ensuring that the final draft is submitted before the viva voce
- providing support and encouragement throughout the EE process
- setting up a clear schedule for reflection sessions and they may set up a dedicated space for your reflections
- reading the final version in conjunction with the viva voce to confirm authenticity
- reporting any suspected misconduct to the IBDP or EE coordinator.

When to do the EE

The timeline of the EE process is up to the school. Some schools start earlier than others but it all needs to be finished by about three months before your Diploma exams.

What you must do

The EE is a substantial piece of work and you must understand the expectations of it and manage your time and workload effectively.

- Having chosen ESS as your subject for the EE think very carefully about the research question and make sure it fits ESS.
- If you are opting to do an EE in world studies with ESS as one of the subjects make sure you have chosen an issue of global significance from the six world studies themes.
- Ensure that you observe the ethical guidelines.
- Know the ESS requirements for an EE.
- Read and make sure you understand the assessment criteria.
- Have a clear plan for the essay.
- Plan ahead – know how, when and where you will find the material and sources you need for the essay.
- Plan a schedule for research and writing and factor in extra time for delays and problems.

- Record sources for your research AS YOU GO; you may not remember them all at the end.
- Meet all internal deadlines set by your school.
- Understand the concepts of academic honesty, including plagiarism and collusion.
- Make sure you have acknowledged all sources of information and ideas in a consistent manner.
- Attend the three mandatory reflection sessions.
- Make the most of the time with your supervisor – go prepared with questions or discussion points.
- Proofread the first draft carefully and make sure it is complete.
- Proofread the final version and make sure it is correctly and consistently referenced.
- Make sure you submit the correct version for assessment.
- Record your reflections from each session on the RPPF and make sure it is signed and dated by you and your supervisor.

More on reflection

Student reflection is a critical part of most of your academic endeavors including the EE. Reflection will highlight the journey you have taken with your EE and how you have changed on that journey.

Reflection requirements in the EE are a bit like the process journal for MYP. There are three mandated reflection sessions in the EE. If there is a reflection space set up, the ideas that feed into these discussions should be noted there.

The first reflection session

The discussion should include:

- initial topic ideas
- possible sources of information / method of data collection
- preliminary research question
- personal thoughts on the topic
- timeline.

The interim reflection session

The discussion may include:

- progress on your thinking
- development of arguments
- questions raised by your research / data collection
- reactions to what you have discovered
- progress on your timeline
- setbacks/problems you have encountered
- strategies to move forwards.

The final reflection session – viva voce

This takes place after the completion and submission of the final essay.

- You may be able to show what you have learnt about the topic, the research process, and your own learning.

- You can share any new questions or issues you have uncovered.

- You may highlight the personal significance of the work you have done.

Writing the extended essay

The EE must be in the correct formal academic style; therefore, you must stick to the following:

- Font: Readable – nothing fancy

- Size: 12 points

- Double spaced

- Pages numbered

- The following MUST NOT appear anywhere on the essay:

  - your name or candidate number

  - the school's name or number.

- Word limit – 4000 words NO MORE:

  - Examiners will not read past 4000 words

  - E-uploading marks the automatic cut-off point for assessment, so stay below it.

Word count

What is included in the word count?

- Introduction
- Main body
- Conclusion
- Quotations
- Footnotes and endnotes (that are not references)

What is not included in the word count?

- Contents page
- Maps, charts, diagrams, annotated diagrams
- Equations, formulas and calculations
- Citations/references

Structuring your essay

The EE must have the following six elements in this order:

- Title page

- Contents page

- Introduction

- Body of the EE

- Conclusion
- References and bibliography

Title page

The title page should include ONLY the following information:

- the title which:
    - is a clear focused statement of your research
    - shows the research topic
    - is NOT phrased as a research question.
- the research question
- the subject the EE is registered for
- the word count.

Contents page

All pages in the EE must be numbered. The contents page must come after the title page and should include reference to all the sections.

Introduction

This section needs to tell the reader:

- about the focus of the essay
- the scope of the research question
- the sources that will be used
- the line of argument.

Main body of the EE

This is the main part of the EE. It contains the research, analysis, discussion and evaluation and should:

- be presented as a reasoned argument
- show what relevant evidence has been discovered
- show where or how the evidence was discovered
- show how the evidence supports the argument

You may choose to divide your work up under sub-headings for greater clarity. Remember however that all information that is important to the argument MUST be in the body and not in the appendices. The examiner does not have to read the appendices so anything in there will not be considered. Avoid appendices as much as you can. If you feel you have to use them, then use them only for the following information:

- Exemplars
- Questionnaires
- Interview questions
- Permission letters (if used)
- Raw or statistical data tables

Conclusion

This must relate to the question posed and should include:

- what has been achieved
- notes on limitations

- questions that have not been resolved
- a final summative conclusion of any conclusions drawn throughout the body of the essay.

References and bibliography

It does not matter which style of referencing you use so long as you are consistent. Make sure you reference as you go – that way you are less likely to forget something. If you are unsure of anything to do with referencing talk to your supervisor or check the IB document *Effective Citing and Referencing*.

How to select the right ESS extended essay

If you choose ESS for your extended essay, your work must have a focus on the **interaction and integration** of natural environmental systems and human societies. You cannot cover an ecological topic if there is no human interaction or effect on humans. The relationship between the natural environment and human societies must be evident, but there does not need to be a 50/50 balance. Make sure that you select a topic in which you are truly interested. Your research question must be open to analytical argument. A narrative or review of published papers is not enough. Do not just describe; you need to evaluate and compare. You also need to support your argument with evidence from your research; you cannot simply rely on secondary sources. It may be that your research question is better fitted in another Diploma subject e.g. Biology or Geography–check this out before you start writing.

Always consider the IB ethical guidelines. You must not consider an EE where pain is inflicted on a living organism, the environment is damaged or human health under threat.

Common errors

- You have not got a research question that focuses on environmental systems AND human societies.
- Your EE is not open to analytical argument – it is only a description or narrative.
- You are unethical or unsafe in your choice.
- Your research question is too broad or unfocused.
- Your research question cannot be answered in 4000 words.
- Your sources are not relevant to the research question.
- You cite websites uncritically.
- You have a one-sided argument and no counterclaims.

Getting the title right

While your research title will be specific to your interests, some of these and a few comments about them may help you in clarifying your ideas.

Having a focused title can set you on the right track for your EE. Here are examples of too broad and focused topics.

Too broad a title	Focused title
Paper recycling	Investigating the environmental impact of paper use at an international school in the Netherlands
Efficiency of world food production	A comparison of a dairy farm in the Netherlands to one in Tanzania.
Impacts of mining on the environment	An investigation of the recovery and restoration after bauxite quarrying in a mine in Western Australia

Some examples of high-scoring EE titles

- How would varying acidic watering solutions, simulating acid rainfall, affect the growth of purple loosestrife (Lythrum salicaria), in an attempt to assess the effectiveness of the acidic nature of southern Ontario precipitation as a control for the invasive growth characteristics of purple loosestrife?

- The potential of residential solar power systems to meet the grid demand in Canberra, Australia

- The impact of urban development on wild bee populations in the Washington DC area

- A soil analysis of the Colorado Mountain pine beetle epidemic

- Sustainable land use in rural post apartheid South Africa: an investigation of the Mhlumeni community in the Onderberg region of Nkomazi Local Municipality in Mpumalanga Province (This is taken from an IB publication *50 Excellent Extended Essays*.)

- To what extent the River Ganga in the Sangam city was affected environmentally amidst the Maha-Kumbh 2013 (This was submitted as a Geography EE but could also have been an ESS EE.)

- Evaluation of water quality in Ramallah district springs

- What impact do soil erosion rates have on the people and habitat in two contrasting areas on the lower slopes of Kilimanjaro, Tanzania?

- Is the decline in wild salmon in Vancouver due to the increase in number of farmed salmon?

- A comparison of a dairy farm in the Netherlands to one in Tanzania

- Which ecosystem within the UK suffered the worst ecological fallout as a result of the Chernobyl disaster?

- An investigation into the soil erosion rates and effect on people and habitat in two areas of the lower slopes of Mt Kilimanjaro, Tanzania

- Comparing inputs, outputs and efficiencies in an organic and non-organic sheep farm in New Zealand

- What are the environmental and social impacts of building the Three Gorges Dam in China? (Although based on secondary data, some siltation experiments were carried out.)

Some EE titles that are too general

All these are far too broad and global or do not meet ESS requirements and cannot be addressed within the EE word limit to meet the criteria. Avoid such general titles:

- Environmental effects of mining

- Efficiency of world food production

- Recycling of paper

- Tourism in Antarctica

- Global dimming (However this could be a good idea if it were experimental and addressing light and/or heat transmission through different levels of atmospheric pollutants and their effect on the growth/germination of plants.)

- Oceans and their coral reefs (However investigating a specific coral reef bleaching or death over time and comparing primary and secondary data would be a better title.)

- The greenhouse effect (Avoid this unless you do specific experiments. If thousands of scientists are working on this, what would you say in 4000 words?)

- The effects of oil spillage on marine life (If you made this specific to one oil spill, or a comparison of two and their clean-up and restoration, you could have a good title.)

- The nuclear winter

Too imbalanced or just not right

- The wisdom of Lynx reintroductions in Colorado (The title suggests the conclusion – that they are not wise – and there is a danger in only putting one side of the argument. It would be better to keep to title open as Lynx reintroductions in Colorado, success rates, methodology and evaluation.)

- Temporary habitats for aspiring Martians (This is perhaps a little too flippant as a title. It was about terraforming but, again, it is too general to score well.)

- The effects of the Chernobyl catastrophe 1986 on Germany and the Bavarian woods 19 years after the event (How could you compare then and now?)

Too vague

- Do the positive consequences of destroying the Amazon rainforest to benefit humans outweigh the negative consequences, or are the negative consequences so critical that there is nothing worth risking them for?

- The study of one key aspect to determine success of a community-based conservation project

- Effect of fertilizer runoff on freshwater ecosystems

- The qualities of water and its impact on plants

Not addressing the ESS subject

- Antimicrobial resistance: a growing ethical dilemma between economics and the environment (This would be better submitted as a Biology EE.)

- Heavy metal contamination in dietary supplements (This is Biology again, but could it be unethical if experimental?)
- Does the frequency of noticeable earthquakes off the west coast of S California and the Gulf of Alaska correspond to Southern Oscillation Index Values? (This is Geography)

Assessment of the extended essay

Your EE is assessed using markbands and your work is awarded marks on a best-fit basis. There are five criteria which are:

Criterion		Maximum marks
Focus and method	A	6
Knowledge and understanding	B	6
Critical thinking	C	12
Formal presentation	D	4
Engagement	E	6

Total 34 marks

Checklist of EE assessment

Does your EE cover all these?

A: Focus and method – 6 maximum

Have you communicated the topic accurately and effectively?

- Have you identified and explained the research topic effectively? ☐
- Have you shown the purpose and focus of the research clearly and appropriately? ☐

Is your research question clearly stated and focused:

- Is the research question clear and addressing an issue of research that is appropriately connected to the discussion in the essay? ☐

Is your methodology of the research complete?

- Is there an appropriate range of relevant source(s) and/or method(s) applied in relation to the topic and research question? ☐
- Is there evidence of effective and informed selection of sources and/or methods? ☐

B: Knowledge and understanding – 6 maximum

Is your knowledge and understanding excellent?

- Is the selection of source materials clearly relevant and appropriate to the research question? ☐
- Is your knowledge of the topic/discipline(s)/issue clear and coherent and are sources used effectively and with understanding? ☐

Is your use of terminology and concepts good?

- Is your use of subject-specific terminology and concepts accurate and consistent? ☐
- Do you demonstrate effective knowledge and understanding? ☐

C: Critical thinking – 12 maximum

Is your research excellent?

- Is your research appropriate to the research question and is its application consistently relevant? ☐

Is your analysis excellent?

- Is the research analysed effectively and clearly focused on the research question? ☐
- Does the inclusion of less relevant research not significantly detract from the quality of the overall analysis? ☐
- Are your conclusions to individual points of analysis effectively supported by the evidence? ☐

Is your discussion/evaluation excellent?

- Do you have an effective and focused reasoned argument developed from the research? ☐
- Do you have a conclusion that reflects the evidence presented? ☐
- Is your reasoned argument well-structured and coherent, with any minor inconsistencies not hindering the strength of the overall argument or the final or summative conclusion? ☐
- Have you critically evaluated the research? ☐

D: Formal presentation – 4 maximum

Is the formal presentation good?

- Is the structure of the essay clearly appropriate in terms of the expected conventions for the topic, argument and subject in which the essay is registered? ☐
- Are your layout considerations present and applied correctly? ☐
- Do the structure and layout support the reading, understanding and evaluation of the EE? ☐

E: Engagement – 6 maximum

Is your engagement excellent?

- Have you made evaluative reflections on decision-making and planning? ☐
- Have you included reference to your capacity to consider actions and ideas in response to setbacks you experienced in the research process? ☐
- Do your reflections communicate a high degree of intellectual and personal engagement with the research focus and process of research? ☐
- Do you demonstrate authenticity, intellectual initiative and/or a creative approach? ☐

Big Questions in ESS

If you have read the ESS guide, you will have seen the Big Questions:

a. What strengths and weaknesses of the systems approach and the use of models have been revealed through this topic?

b. To what extent have the solutions emerging from this topic been directed at *preventing* environmental impacts, *limiting* the extent of the environmental impacts, or *restoring* systems in which environmental impacts have already occurred?

c. What value systems can you identify at play in the causes and approaches to resolving the issues addressed in this topic?

d. How does your own value system compare with others you have encountered in the context of issues raised in this topic?

e. How are the issues addressed in this topic of relevance to sustainability or sustainable development?

f. In what ways might the solutions explored in this topic alter your predictions for the state of human societies and the biosphere some decades from now?

But what have you done with them?

The idea of the big questions is that they shape your learning in ESS by asking you to think about the whole picture – holistically – and to make links across topics and concepts.

What does that mean to you?

Because they are not examined as such, you may have not taken much notice of them but they are helpful if you use them to revise.

What you need to do is to make the questions more specific to the topic you are revising. Then you could use them as practice questions for Paper 2 Section B part (c) essay-style questions.

So here goes in making the big questions specific to topics.

a. Evaluate the use of models to assess sustainability.

b. Explain what are the main impacts of humans on flows of energy and materials in the biosphere.

c. To what extent are our solutions to ocean pollution mostly aimed at prevention, limitation or restoration?

d. Discuss in what ways your environmental value system influences your attitude towards conservation of species.

e. To what extent could food production become more sustainable?

f. Discuss your predictions for the state of a world without oil from fossil fuels.

Can you see that the idea of the big question is still there in this list above but it is related now to a topic?

> Select one Big Question and rewrite it with relevance to any ESS topic.
>
> Then write an essay plan for the question and give it a mark out of 9. (refer back to the markscheme on page 160)
>
> Repeat for as many questions as you wish.
>
> This is good revision and helps you make connections between topics, which is what you need to do in the exams.

The Lorax

Have you heard of *The Lorax* by Dr Seuss? It is a book and a film and should be easy to find.

If you have not read it or seen it before, get a copy and use it for ESS revision. Yes! It is great for revising this course.

Dr Seuss wrote *The Lorax* in 1971, perhaps inspired by the ecological disasters of the 1960s and 1970s to do something about it. (Rachel Carson wrote *Silent Spring* in 1962.) The Lorax is a warning to us about upsetting the balance of the ecosystem and the dangers of clear-felling forests, pollution and industrialization. It is as relevant now as when it was written and a story for children and adults.

The story is told in rhyme by the Once-ler who lives alone and is faceless and bodiless as he narrates it to a boy.

Long ago, the Once-ler finds a place filled with Truffula Trees, Swomee-Swans, Brown Bar-ba-loots, and Humming-Fishes. It was so beautiful and empty of people that he stopped there and started chopping down the trees to make Thneeds. It is not clear what a Thneed is but his marketing makes everyone want one. As more trees are felled to turn into Thneeds, and none replaced, the Swomee-Swans, Brown Bar-ba-loots, and Humming-Fishes die or leave. The Lorax – yellow and furry – keeps warning the Once-ler and speaks out for the trees but the Once-ler ignores him. In the end, even the Lorax leaves and the once beautiful forest is a scorched earth where nothing grows. Left behind is a rock on which the word 'unless' is written.

The hopeful note is that at the end the Once-ler gives the boy the single seed he has saved of a Truffula Tree and it seems that it is all up to him now to restore the once wonderful forest.

Watch or read the Lorax story.

Now the revision part.

1. Define and give two examples of non-renewable natural capital.

2. What is the difference between natural capital and natural income?

3. Define sustainable yield.

4. **a.** What should the Once-ler have done when he exceeded the sustainable yield of the forest?

    b. How is this an example of the Tragedy of the Commons?

5. If you were the Once-ler, what would you have done differently to sustain the forest?

6. A Truffula forest contains 6000 trees. Each year on average, 500 young trees start to grow and survive to maturity after 30 years. The trees can be cut when they are mature at 30 years. What would be a sustainable yield for this forest?

7. What are other forms of natural capital that might also be used or lost when the Once-ler cuts down the trees?

8. **a.** Draw a flow diagram of the system of the forest.

    b. Identify a positive and a negative feedback mechanism in the forest on your flow diagram.

9. Write your own ending to the Lorax story starting after the word 'Unless'.

10. Relate the Lorax story to a case study that you know.

What other books or films do you know of that could be used to revise in ESS?

Examples might be *2012*, *Avatar*, *Wall-E*, *The Day After Tomorrow*, *Happy Feet* and *Erin Brokovich*. This is not a call to spend your revision days watching movies but think about the messages and ESS concepts in the movies you have seen.

Resources

By the time you start revising – remember that is what you do when you revisit what you have studied – you should have covered all the ESS course, so there should be nothing new to surprise you. Use your textbook as you should be familiar with it by now, and study your notes and lab reports.

But sometimes you may want to expand on a topic, check your understanding or look up some examples and case studies.

Others have been through the course and the exams too – about 10,000 a year – and sometimes they and their teachers put ESS resources on the web. If you need a change of revision tactics or just a change, try some of these. Just remember to cite your sources at all times.

These are revision sites but BE CAREFUL as at the time of writing they cover the old course, not the one with first exams in May 2017.

http://envirohome.wikispaces.com/IB+ESS+EXAM+NOTES – revision site

http://sciencebitz.com/ – from Mr Nigel Gardner – great site for info.

http://www.i-study.co.uk/IB_ES/Environmental_systems_homepage.html

http://www.bend.k12.or.us/education/components/docmgr/default.php?sectiondetailid=4474 from Paul Hutter at Bend Senior High School, Oregon

On DDT

from Jin Young Park http://www.personal.psu.edu/users/j/o/jop5349/Assignment7.html

On ecological footprints

http://www.footprintnetwork.org and http://footprint.wwf.org.uk/

On the water cycle and ocean currents

http://science.nasa.gov/earth-science/oceanography/ocean-earth-system/ocean-water-cycle/

And, of course, YouTube videos and films. But be careful here too – you can waste a lot of time searching for a useful clip and get distracted very easily.

On water

http://www.bluegold-worldwaterwars.com/ Documentary about water security.

http://www.tappedthemovie.com/ Documentary about the influence of the bottled water industry.

http://www.ted.com/talks/anupam_mishra_the_ancient_ingenuity_of_water_harvesting.html The Global Water Partnership and a discussion of water security.

http://endoftheline.com Documentary about overfishing around the globe.

http://www.thecovemovie.com/ Documentary about dolphin hunting in Japan.

http://movies.nytimes.com/2005/08/03/movies/03darw.html?_r=0

Darwin's Nightmare (2005), showing the effects of an introduction of an invasive species, the Nile Perch, to Tanzania.

http://www.unep-wcmc.org/resources-and-data

http://www.ted.com/themes/ocean_stories.html

http://www.ted.com/themes/a_taste_of_mission_blue_voyage.html

http://vimeo.com/25563376 Trailer for the environmental documentary *Midway: Message from the Gyre,* about the oceanic plastic pollution problem and its impact on wildlife.

On food

http://www.freshthemovie.com/

FRESH The Movie – a documentary about food production systems in the United States.

http://video.pbs.org/video/1402965302/

Trailer for *FOOD Inc.*, documentary about food production systems, Monsanto, and the industrialization of food production in the US.

On soil

http://www.bbc.com/future/story/20120209-mud-mud-glorious-vanishing-mud

Games

http://www.catchmentdetox.net.au/

Index

British Library Cataloguing in Publication Data
Data available

978-0-19-836669-0

5 7 9 10 8 6 4

Paper used in the production of this book is a natural, recyclable product made from wood grown in sustainable forests. The manufacturing process conforms to the environmental regulations of the country of origin.

Printed in India

Acknowledgements
The authors and the publisher are grateful for permission to reprint extracts from the following copyright material:

American Society of Agronomy Inc: Instructional diagram for determining soil texture by feel (modified), republished with the permission of American Society of Agronomy Inc, from 'A flow diagram for teaching texture by feel analysis', S.J. Thien, *Journal of Agronomic Education*, 8, 1979; permission conveyed through Copyright Clearance Center, Inc.

The International Baccalaureate Organization: Extracts from 'Environmental systems and societies guide', first examinations 2017, published 2015; Assessment Objectives from 'Environmental systems and societies guide', first examinations 2017, published 2015; Command Terms from 'Environmental systems and societies guide', first examinations 2017, published 2015; Examination questions from past papers, May 2009 Paper 2, May 2010 Paper 2, May 2011, Paper 2. This work has been developed independently from and is not endorsed by the International Baccalaureate (IB).

Professor William E. Rees: Definition of 'ecological footprint', from http://www.wwf.org.au/our_work/people_and_the_environment/human_footprint/ecological_footprint/.

The publishers would like to thank the following for permissions to use their photographs:

Cover image: Shutterstock; P2: Bart Coenders/iStock; P17: Ttphoto/Shutterstock; P31tl: Jeffrey Bruno/Shutterstock; P31tr: visualdestination/Shutterstock; P31ml: gnomeandi/Shutterstock; P31bl: Seth LaGrange/Shutterstock; P31br: Denis Burdin/Shutterstock; P33: 2015 Global Footprint Network. www.footprintnetwork.org.; P57b: photong/Shutterstock; P58: Corbis; P67: Semmick Photo/Shutterstock; P70: Corbis; P71: Yuri Kravchenko/Shutterstock; P80t: Russell Millner/Alamy Stock Photo; P80m: Radiokafka/Shutterstock; P80b: Joseph Sohm/Shutterstock; P81: Koraysa/Shutterstock; P87: John Prieto/The Denver Post/Getty Images; P89: BlueRingMedia/Shutterstock; P99l: Utopia_88/Shutterstock; P99r: Ethan Daniels/Shutterstock; P100tl: denniro/Shuttersock; P100bl: Alamy Stock Photo; P100br: Matthew Heinrichs/Alamy Stock Photo; P104: Antonov Roman/Shutterstock; P105: Pongthorn S/Shutterstock; P112l: kryzhov/Shutterstock; P112m: KreativKolors/Shutterstock; P112r: Marko Cermak/Shutterstock; P118l: Utopia_88/Shutterstock; P118r: Dudarev Mikhail/Shutterstock; P119tl: Apinan/Shutterstock; P119tr: smereka/Shutterstock; P119ml: ChiccoDodiFC/Shutterstock; P119mr: jeremy sutton-hibbert/Alamy Stock Photo; P119bl: Bildagentur Zoonar GmbH/Shutterstock; P119br: Lev Kropotov/Shutterstock; P121l: lapon pinta/Shutterstock; P121r: Richard Cavalleri/Shutterstock; P124: Bartek Zyczynski/Shutterstock; P125: George Dolgikh/Shutterstock; P126: Russell Gillman/Alamy Stock Photo; P131t: wavebreakmedia/Shutterstock; P131b: wavebreakmedia/Shutterstock; P141: U.S. Coast Guard/Getty Images; P142: George Konig/Keystone Features/Getty Images; P147: Wikimedia Commons/Public Domain; P184: AF archive/Alamy Stock Photo;

Artwork by OUP and Six Red Marbles.

Although we have made every effort to trace and contact all copyright holders before publication this has not been possible in all cases. If notified, the publisher will rectify any errors or omissions at the earliest opportunity.